Organic Chemistry Practicals and

Important Reagents

:: Author ::

Dr. Darshan V. Chaudhary

PUBLISHED BY

Hemchandracharya International Publishing House
H.Q. At & Po. Chaveli., Ta- Chansma,
Dist- Patan, North Gujarat, India, Asia.
www.iphouseindia.com

First Publication: 1st March, 2016

Copyright: Author

(c) **Dr. Darshan V. Chaudhary**

ISBN:- 978-1-53041-719-3

Price: Rs.800/- INDIA

   $ 15 OUTSIDE INDIA

PUBLISHED BY

Hemchandracharya International Publishing House
H.Q. At & Po. Chaveli., Ta- Chansma,
Dist- Patan, North Gujarat, India, Asia.
www.iphouseindia.com

PREFACE

The goal of this manual is to provide a satisfactory, elementary attention of experimental organic chemistry in a one-semester laboratory course. It will therefore oblige the requirements of non-chemistry majors (pharmacy, biology, environmental science, agriculture etc.) needful an initial development in experimental organic chemistry. Chemistry majors may also benefit from this manual if it is used as an introductory course to be followed by more advanced laboratory techniques and syntheses.

The experiments are intended to familiarize students with methods generally used in the organic laboratory for identification, purification and separation of organic compounds. In addition simple preparative experiments are involved to demonstrate the chemical behavior among different classes of organic compounds. The accessibility and rate of the chemicals were taken into consideration when choosing the experiments.

The manual is separated into two parts. Section I cover the basic techniques while in Section II simple, one-step syntheses and chemical tests for the main classes of organic compounds are presented. Each experiment is followed by a report sheet. For suitability, the estimated time for completion of every experiment and appropriate hazard symbols are indicated in the experimental section.

ACKNOWLEDGMENT

I express my heartfelt thanks to Dr.Pranav Srivastav, Professor, Chemistry Department, Gujarat University, Ahmedabad, India, for his constant guidance across my research Work and without the same platform I would not be able to compile this book. I am very thankful to my other colleague contributor Mr.Edvin Pithawala for critical evaluation of each chapter across the book. I would like to express my gratitude to my family members especially to my parents for their love affection and care and last but not the least to my beloved Reema for her everlasting love, motivation and sacrifice for the time taken in compiling this book.

I am grateful to publisher for their concern, efforts and encouragement, especially for their excellent cooperation in the task of preparing and publishing this book.

- Dr. Darshan Chaudhary

TABLE OF CONTENTS

GENERAL INFORMATION

SAFETY PRECAUTIONS:

Safety in the laboratory can be achieved only when common sense and consideration for othersprevails, while following the basic regulations outlined below:

1. Smoking is not allowed in the laboratory.
2. Report all accidents immediately to the instructor.
3. Know the location of fire extinguishers and how to use them.
4. If any person has hair or clothing on fire, as a first step, lie down on the floor and use a blanket, coat or anything available to smother the flames. Get help immediately.
5. Experiments should never be left unattended.
6. Never taste any solid or liquid chemical.
7. When smelling a substance do not hold your face directly over the container.
8. Most organic substances are hazardous to health; so avoid breathing and skin contact as much as possible.
9. Working with toxic, lachrymatory or irritating chemicals must be conducted in fume hoods. In some cases a trap must be used to prevent hazardous gases from escaping into the laboratory atmosphere.
10. If acids or corrosive chemicals are spilled on your skin, wash with plenty of cold water then consult your instructor.
11. Do not point your test tube at your neighbour or yourself when heating substances.
12. Most organic solvents are flammable, so never heat a flammable substance with a direct flame. A hot water bath is used instead.
13. Always wear a laboratory coat.
14. If acid or base is spilled on your clothing, bench or floor

wash thoroughly with water, then neutralize with dilute ammonium hydroxide or acetic acid respectively and inform your instructor.

15. It is advisable to wear safety glasses in the laboratory.

16. Always wash your hands with soap and water on leaving thelaboratory.

HAZARD WARNING SYMBOLS:

Corrosive   Harmful   Oxidizing

Explosive

LABORATORY INSTRUCTIONS:

The laboratory rules given below should be strictly followed.

1. Throw all solids to be discarded into a waste paper basket. Never throw matches, filter paper, broken glass or any insoluble chemicals into the sink. Organic liquids should not be poured into the sink either. They are collected in special residue bottles.

2. Read the label twice before using a reagent bottle.

3. The reagent bottles on the side shelves should not be carried to your bench. Pour the solution you require into a

2

beaker and at your bench measure the amount you need. This prevents crowding at the side benches.

4. Never return chemicals to the stock bottles.

5. Do not lay the stopper of a bottle down, as impurities will be picked up and contaminate the stock solution. Also, close reagent bottles immediately after use.

6. At the end of each laboratory period, leave your glassware clean, and the top of your bench clean and dry.

7. Study the experiment and read the directions carefully before coming to the laboratory. Know what you are going to do and why, and be ready to answer questions on the experiment.

8. Unless otherwise told, each student is required to work the experiment alone. Copying of data and working together is absolutely forbidden.

9. The following items have to be brought into every laboratoryperiod:
    - *a.* laboratory manual,
    - *b.* laboratory coat,
    - *c.* matches,
    - *d.* dish towel,
    - *e.* sponge for desk cleaning.

REPORTING RESULTS:

1. Students are required to record all data and observations collected during the experiment on the report sheets provided for that purpose.

2. Numbers resulting from measurements or calculations must be accompanied by correct units.

3. Yield calculations: The following reaction is given as an example:

benzoic acid — Methanol — Methylbenzoate
(mw 122) — (mw 32) — (mw 136)

From the stoichiometry of the balanced equation, one mole of benzoic acid reacts with one mole of methanol to give one mole of methylbenzoate assuming 100% reaction. If a student actually weighed 12.2 g (0.12mol) benzoic acid and measured 20 mL (15.8 g, 0.49mol) of methanol, then methanol is present in excess while the amount of benzoic acid (0.12mol) limits the amount of methylbenzoate produced. Benzoic acid is therefore the limiting reactant. Only 0.12mol (13.6 g) of the methyl benzoate can theoretically be produced.

The actual yield is the amount of product obtained experimentally. It may differ from one student to another. The percentage yield is given as:

$$\text{percentage yield} = \frac{\text{actual yield}}{\text{theoretical yield}} \times 100$$

If the student actually obtained 9.5 g of product then his percentage yield would be (9.5 /13.6 x 100) = 70%.

4. At the end of each laboratory period, you should submit your product to the instructor with a label that shows your name, the name of the product, and any relevant data.

5. Questions in the report sheet should be answered in the laboratory and handed in at the end of each period.

COMMON LABORATORY TECHNIQUES

HANDLING OF GLASSWARE:

Dirty glassware may be cleaned with soap and water using a brush. However, glassware which has persistent stains from organic substances requires soaking in chromic acid cleaning solution. This mixture has to be used carefully as it is very corrosive.

Glass tubing with unpolished ends is a hazard since it can cause serious cuts when trying to insert it into a cork. Therefore, only glass tubing with polished ends must be used. When forcing glass tubing into a cork, grasp it as close as possible to the cork and be careful not to break it.

Quick fit glass joints should always be lubricated with a suitable lubricant (grease). A thin film of grease is applied to the joints to provide an air-tight seal and to prevent the joints from being stuck together. There should be no excess grease extending inside the apparatus as it might contaminate the reaction mixture. It is also recommended that old grease be wiped off with a piece of tissue paper before applying a new film.

HEAT SOURCES:

There are various heating devices in the laboratory. The Bunsen burner and water bath are the most commonly used. A limitation of the Bunsen burner is that it should not be used directly for heating flammable solvents.

Flammable and volatile liquids are heated in a water bath when temperatures under 100 are required. If an electrical steam bath is not available, a large beaker filled with water may be used instead. It is heated to boiling with a Bunsen burner and the flame extinguished before heating the flammable liquid in the bath. Two disadvantages of the water bath are:

a. The water in the bath evaporates when long periods of heating arerequired.

b. It is not recommended when anhydrous conditions are necessary.

Other heat sources also used in laboratories are hot plates, oil baths and heating mantles.

BUMPING:

When heated, liquids may get superheated, i.e. reach a temperature above their boiling points. A superheated liquid exists in a metastable state and tends to regain equilibrium through sudden vigorous vaporization which carries hot liquid along with escaping vapors. This phenomenon, called bumping, may be prevented by continuous stirring to ensure homogenous and steady heating of the liquid or by the use of boiling stones which achieve a similar effect through formation of bubbles.

FILTRATION:

Filtration is used whenever an insoluble solid is to be separated from a liquid. Simple gravity filtration (usually hot filtration) is employed to remove insoluble solid impurities from a liquid, while suction filtration (usually cold filtration) is used to collect a desired solid or crystalline product. (Figures 1 and 2)

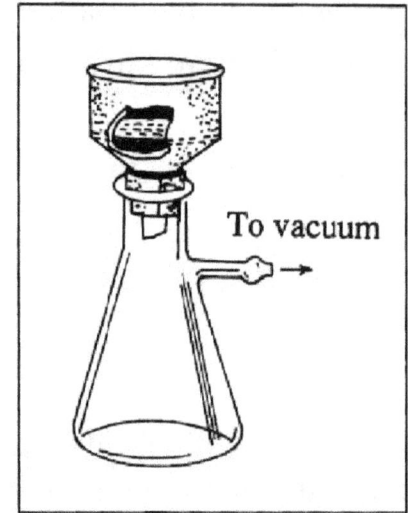

Figure 1.Simple filtration.Figure 2. Suction filtration.

DECOLOURIZATION:

Decolourization is the removal of coloured impurities from a solution. This is achieved by the addition of activated charcoal to the solution and mixing thoroughly. If charcoal is added to a cold solution, the solution is first brought to a boil before hot filtration. When however it is added to a hot solution, the flask should be removed from the heat source before the addition, otherwise bumping will occur. Charcoal is finally removed by filtration leaving an almost colourless solution.

DRYING AND DRYING AGENTS:

The process of drying, if applied to a solid substance (e.g., after recrystallization), is intended to remove residual solvent (organic or water) adhering to the solid particles or crystals. This is usually done by air drying (spreading over a sheet of paper) and/or heating in an oven to enhance evaporation of the solvent.

Drying of an organic liquid, however, involves the removal

of traces of water (moisture) using chemical drying agents. Such cases are encountered in extraction where the organic phase is in direct contact with the aqueous phase. After separating the layers, traces of water in the organic phase are removed by the addition of a suitable drying agent.

One class of drying agents are the anhydrous inorganic salts which combine with water to form hydrates. Since these drying agents are insoluble in common organic solvents, they are easily removed by decantation or filtration once drying is complete. Some common examples are: calcium chloride, magnesium sulphate, sodium sulphate, sodium hydroxide and potassium hydroxide.

REFLUX:

The technique of refluxing is commonly used when it is necessary to heat a reaction in order to bring it to completion in a reasonable time span. Areflux condenser is used to minimize loss, through evaporation, of volatile reactants, products or solvent by allowing the vapours to recondense and return to the reaction vessel (Figure 3).

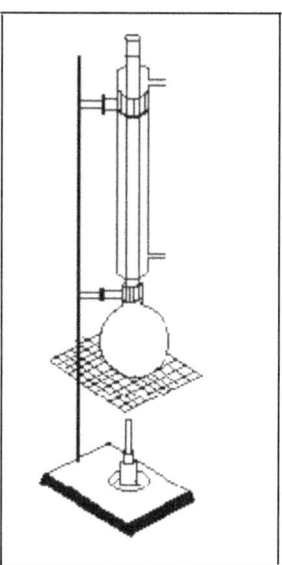

Figure 3. Reflux apparatus.

TRAPS:

Traps are used to prevent hazardous gases and vapours formed in a reaction from escaping into the laboratory atmosphere. Even if such reactions are carried out inside a fume hood, a trap is recommended. In the preparation of n-butyl bromide, for example, a trap is used to absorb escaping hydrogen bromide gas. The liquid in the trap consists of dilute sodium hydroxide solution. The gas is led through proper glass tubing into the trap which should be vented to prevent back suction. Alternatively, one may assemble a trap using a beaker and a filter funnel over the absorbing liquid (Figure 4).

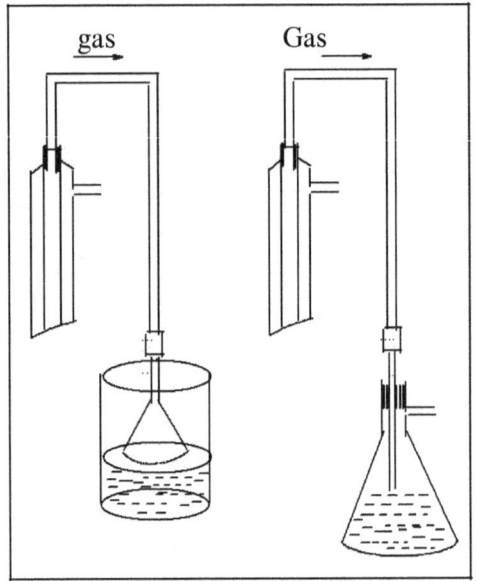

Figure 4.Gas absorption trap.

COMMON LABORATORY EQUIPMENT:

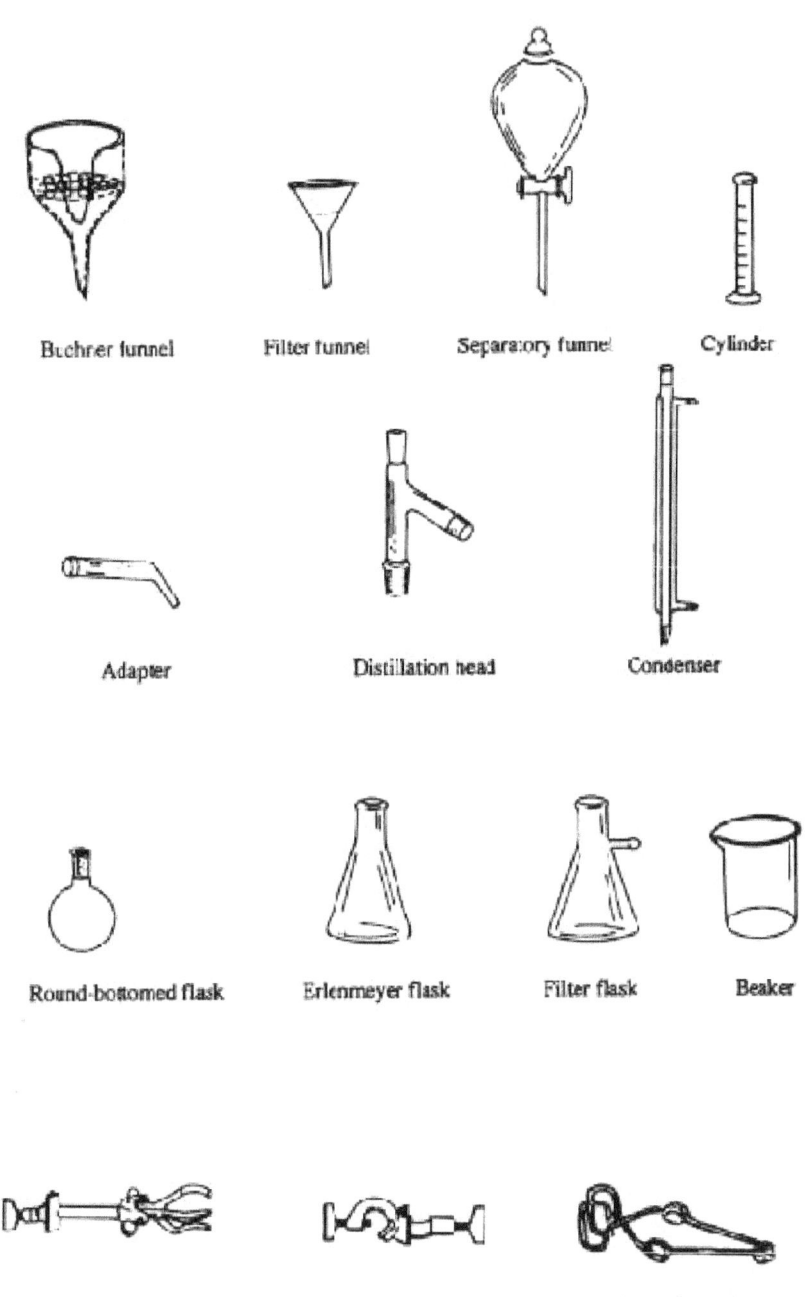

Buchner funnel   Filter funnel   Separatory funnel   Cylinder

Adapter   Distillation head   Condenser

Round-bottomed flask   Erlenmeyer flask   Filter flask   Beaker

Clamp   Clamp holder   Test tube holder

MELTING POINTS
Identity and Purity of Solid Organic Compounds

INTRODUCTION:

The melting point of a solid is the temperature at which transition from solid to liquid occurs at atmospheric pressure; or the temperature at which solid and liquid phases are in equilibrium at a pressure of one atmosphere. The melting point is practically unaffected by changes in external pressure, making it a convenient physical constant for the identification of solids.

Many organic compounds are solids at room temperature as a result of strong intermolecular forces which hold the individual molecules together in a crystal lattice. The nature and strength of these intermolecular forces are responsible for the observed differences in melting point. In general, if the forces are strong, the melting point will be high, and if they are relatively weak, the melting point will be low.

A pure solid has a sharp melting point and will melt within a narrow range of 1-2°C. Soluble impurities affect the melting point of a solid in the following manner:

a. Lower the melting point of the substance, with the upper limit considerably below the true melting point. The presence of an impurity in the molten compound reduces its vapour pressure thus lowering the melting point of the compound (Figure 5a). The greater the amount of impurity, the greater is the melting point depression (Figure 5b).

b. Broaden the melting point range. Depending on the amount of impurity, the melting process may extend over

11

a range of 2-20°C or more. Insoluble impurities (*e.g.*, glass, sand ...*etc.*) do not affect the melting point or the melting point range.

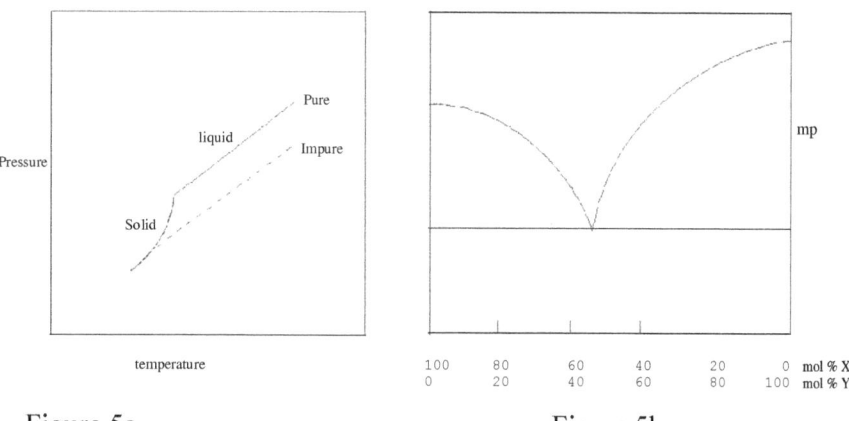

Figure 5a.   Figure 5b.

Vapour pressure-temperature diagram.   Temperature-composition diagram.

Mixtures melting points can be used in the following manner to determine whether two compounds are the same or different even though they have similar melting points. Assuming that a given organic compound (A) melts sharply at 120°C, and benzoic acid (compound B) also has a melting point of 120°C. Is compound (A) benzoic acid or a different compound?

If compound (A) is benzoic acid, then a mixture melting point of (A) and (B) will melt sharply at 120°C, *i.e.* the same as each individual compound alone. If, on the other hand, compound (A) is not benzoic acid, then the mixture melting point of (A) and (B) will be lowered and the melting range will be broadened. Since they are different compounds, each behaves as an impurity in the other.

GENERAL PROCEDURE:

Apparatus: A simple device for determining melting points is shown inFigure 6. It consists of a thermometer fitted through a cork and suspended into a long-necked flask which is three quarters filled with a high boiling and stable liquid like paraffin oil, di-butyl phthalate or silicon oil.

The thermometer bulb should be about 1 cm above the bottom of the flask. The sample in the capillary tube is fastened to the thermometer with a rubber band placed above the level of the oil. The capillary tube should be close to and on a level with the thermometer bulb.

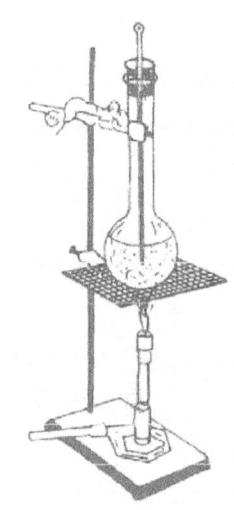

Figure6.Melting point apparatus.

Capillary melting point tubes are about 6-7 cm in length and 1 mm in diameter. They are sealed by rotating one end of the capillary tube in the edge of a small hot flame. The dry solid is ground to a fine powder on a piece of paper with a spatula. The open end of the capillary is then pushed into the powder which is forced down the capillary tube by gently tapping the closed end on the bench top. This is repeated several times until the solid is densely packed at the bottom of the tube to a height of 2-3 mm.

Procedure: To determine the melting point of a solid, a small amount of the powdered substance is introduced into a capillary tube which is then attached to a thermometer and placed in the oil bath. The bath is heated rapidly to within 20°C of the expected melting point then slowly, and at a constant rate of 2-3 degrees per minute, close to the melting point. The temperature at which the solid begins to melt, and that at which it is completely liquid, is recorded as the melting point range of that substance.

The melting point range is affected by a number of factors in addition to that of purity. Particle size, amount of material used, density of packing in the capillary tube, thickness of the capillary tube and the rate of heating of the liquid bath, are all factors that should be carefully considered to ensure an accurate melting point. The rate of heating is the most critical factoraffecting   experimental results, and should be carefully monitored, particularly close to the expected melting point.

OBJECTIVES:
1. Determining the melting point of a pure organic solid.
2. Determining the melting point of an impure organic solid (mixture).
3. Identifying an unknown from its melting point.

Having done this experiment you will have seen the effect of an impurity on the melting point of a solid substance and theuse of the melting point in characterizing organic solids.

EXPERIMENTAL

MATERIALS NEEDED	Glassware: Long-necked round-bottomed flask (50-mL), cork stopper with a hole, thermometer, wire gauze, Bunsen burner, clamp, clamp holder, stand, capillary melting point tubes, rubber band and spatula. Chemicals: 40 mL di-n-butylphthalate (oil bath), 0.3 g of each of the compounds: urea, cinnamic acid, salicylic acid, benzanilide, adipic acid, citric acid, sulfanilamide, acetanilide, mandelic acid, 2-naphthol, benzoic acid, benzoin, maleic acid, and p-toluic acid.

DETERMINATION OF MELTING POINTS OF PURE COMPOUNDS:

Obtain a small amount (about 0.1 g) of each of the two solid compounds from one of the following pairs and determine the melting point of each.

1. Urea and Cinnamic acid.   2. Salicylic acid and benzanilide.

3. Adipic acid and Citric acid.   4. Benzanilide and Sulfanilamide.

You may attach two capillaries simultaneously to the thermometer, so as to determine the melting point of two samples in a single run.

MIXTURE MELTING POINTS:

Make three mixtures of the previously selected pair of compounds by thoroughly mixing small amounts of the components in the approximate proportions of 1:4, 1:1 and 4:1. Determine their melting points and use the midpoints to make a plot of melting point versus composition in your report sheet.

IDENTIFICATION OF AN UNKNOWN:

Obtain an unknown (from instructor) and determine its melting point as before. To save time make an initial rough determination of the melting point by rapid heating. Once the bath temperature has dropped to about 15-20 degrees below the approximate melting point make a more accurate measurement using a fresh sample and slow heating (about 2 degrees / minute). Now refer to the table below and make a list of possible compounds. These are compounds that have melting points 5 degrees above or below the melting point of the unknown.

To finally identify the unknown, make mixtures of the unknown with each of the suspected compounds (one at a time), and determine the melting point of each mixture. In one case only, will there be no depression of the melting point.

Table 1. Melting points of some organic compounds.

Compound	mp (^{o}C)	Compound	mp (^{o}C)
Acetanilide	114	Maleic acid	135
Mandelic acid	117	Adipic acid	152
2-Naphthol	121	Citric acid	154

Benzoic acid	122	Salicylic acid	158
Urea	132	Benzanilide	161
Cinnamic acid	133	Sulfanilamide	165
Benzoin	133	p-Toluic acid	182

QUESTIONS:

1) What two effects do impurities have on the melting point

16

of an organic compound?

2) For what two purposes are melting points routinely used?

3) What effects on the measured melting point would you expect in each of the following cases:

a) Presence of pieces of glass in the sample.

b) Presence of solvent within the crystals.

c) Rapid heating during melting point determinations.

d) Using too large a sample when determining the melting point?

BOILING POINTS AND DISTILLATION

Identity and Purity of Liquid Organic Compounds
Distillation as a Method for the Separation of Liquids

INTRODUCTION:

If a liquid is kept in a sealed container, equilibrium is eventually established between the liquid and gaseous molecules. The pressure exerted by these gaseous molecules is called the vapour pressure and it increases with increasing temperature of the liquid (Figure 7).

The boiling point of a liquid is defined as the temperature at which the vapour pressure of the liquid equals the external pressure (usually 1 atmosphere). It is also defined as the temperature at which vapour and liquid are in equilibrium at a given pressure.

The boiling point, like the melting point, is a physical constant and may be used to identify unknown organic liquids.

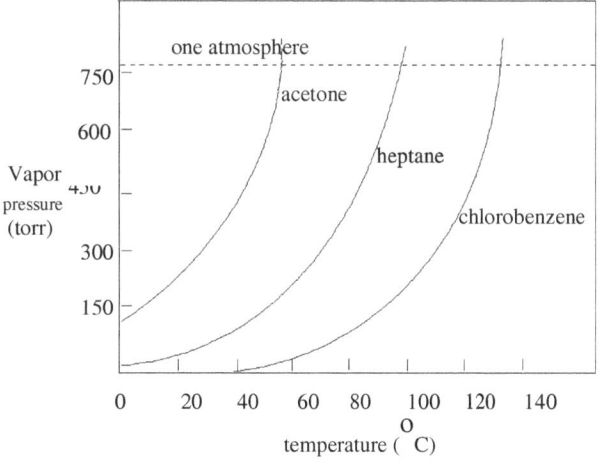

Figure 7. Vapor pressure-temperature curves.

Boiling Points of Pure Liquids: The boiling point of a pure liquid dependson the following variables:

a) Nature of intermolecular attractive forces: H-bonding, dipole-dipole, or London forces.

$CH_3CHCHCHCHCH_2$ —OH $CH_3CH_2CH_2$—O—$CH_2CH_2CH_3$

bp 157 (H-Bonding,)   bp 57 (no H-bonding,)

b) Molar mass: boiling point increases with molar mass.

c) Shape of molecules: among isomeric compounds, having the same functional group, straight chain molecules have higher boiling points than the corresponding branched ones (less molecular surface area).

$CH_3CH_2CH_2CH_2CH_2CH_3$

(bp 68)

$$CH_3-\overset{\displaystyle CH_3}{\underset{\displaystyle CH_3}{C}}-CH_2CH_3$$

(bp 50)

Boiling Points of Solutions: For solutions, the boiling point is defined as the temperature at which the total vapour pressure is equal to the external (atmospheric) pressure. The effect of any solute, A, on the boiling point of a liquid B, will depend on the nature of A (Figure 8). If A is less volatile than B, then the total vapour pressure of the solution is lower at any given temperature, and its boiling point is higher than that of pure B (*e.g.*, a solution of sugar in water). If, on other hand, solute A is more volatile than B, then the total vapour pressure of the solution, at any given temperature, is higher and its boiling point is lower than that of pure B (*e.g.*, a solution of acetone and water).

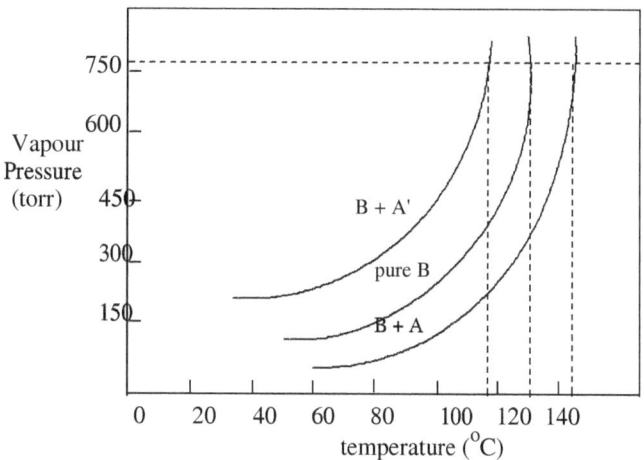

Figure 8. Vapour pressure-temperature diagram for a pure liquid and differentsolutions.

The behaviour of a solution of two miscible liquids, A and B, is best explained by referring to Raoult's law which states that the partial pressure of liquid A (p_A) in a mixture is equal to the vapour pressure of pure liquid A (P^o_A) multiplied by the mole fraction of A in the mixture (X_A). The same applies to liquid B. Therefore:

$$p_A = X_A . P^o_A \qquad \text{and} \qquad p_B = X_B . P^o_B$$

From Dalton's law, the total vapour pressures of the solution (P_T) are the sum of the partial pressures of A and B:

$$P_T = p_A + p_B \quad X_A . P^o_A + X_B . P^o_B \qquad (X_A + X_B = 1)$$

A vapour pressure-composition diagram for the ideal two-component mixture is shown below (Figure 9).

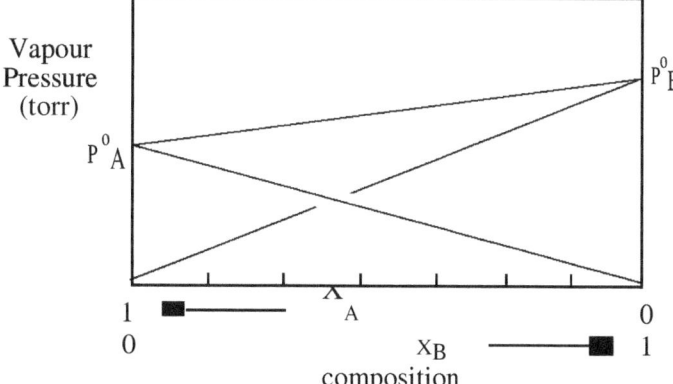

Figure 9. Vapour pressure-composition diagram for two miscible liquids

20

A solution of A and B will boil when the total vapour pressure (P_T) equals the external pressure. This occurs at a temperature which is intermediate between the boiling points of the two pure liquids (lower curve in Figure 10).

In order to understand the separation of a pair of miscible liquids A and B by simple and fractional distillation, a boiling point diagram is helpful (Figure 10).

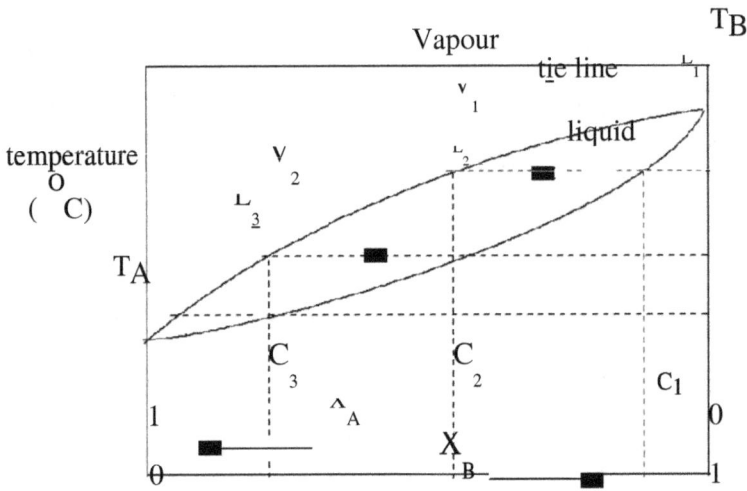

Figure 10.Temperature-composition diagram of two miscible liquids.

This diagram shows the temperature at which mixtures of A and B of various compositions boil (lower curve). The composition of the vapour in equilibrium with the liquid is given by the tie line connecting the liquid and vapour curves. It is clear from Figure 10 that the vapour will always be richer than the liquid in the more volatile component. This makes sense, since the molecules of the component with the higher vapour pressure (more volatile) will escape more readily, and thus be in higher proportion in the vapour phase.

21

Simple and Fractional Distillation: Distillation is the process of heating aliquid to its boiling point, condensing the vapor by cooling, and collecting the liquid distillate. It is a technique for the purification of liquids and for the separation of liquid mixtures. The principles involved in distillation (simple/fractional) may be explained by referring to Figure 10.

If a liquid mixture of A and B with composition C_1 ($X_A = 0.2$) is heated to boiling (L_1), then the vapour in equilibrium with it (V_1) will have the composition C_2 ($X_A = 0.4$), i.e., the vapour will contain more of the volatile component A, than the original liquid. If this vapour is condensed (L_2) and redistilled, the distillate ($V_2 \rightarrow L_3$) will be much richer in A (composition C_3). As the distillation progresses, the mixture will gradually have less ofthe more volatile component and its boiling point will gradually rise. Consequently the distillate will contain a continually decreasing proportion of the more volatile component until finally all has been collected and the less volatile component is left as a residue.

In practice, separation of a liquid mixture into its components by a single distillation (simple distillation) is possible only when the boiling points of the components are 80 degrees or more apart. For mixtures of liquids having boiling points much less than 80 degrees apart, separation can be achieved only by fractional distillation. Such a distillation is equivalent to several repeated simple distillations. It uses a fractionating column which provides a large surface area for continuous heat exchange between the hot ascending vapour and the cooler descending liquid, thus resulting in a series of evaporations and condensations leading to separation of the two components.

Vacuum distillation is a technique for the distillation of high boiling liquids, and for compounds that decompose at atmospheric pressure. At the low pressures employed, those compounds distil at much lower temperatures.

GENERAL PROCEDURE:

A typical set-up for simple and fractional distillation is given in Figures 11and 12, respectively.

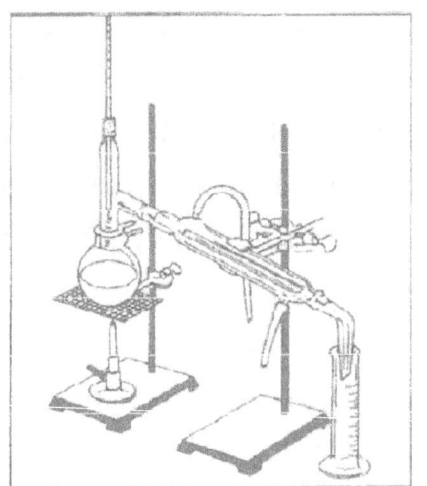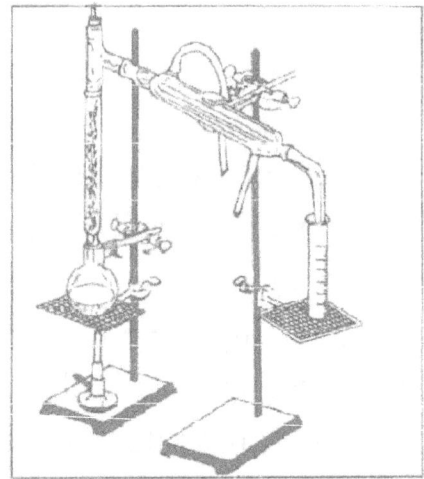

Figure11. Simple distillation apparatusFigure12.Fractional distillation apparatus.

When carrying out a distillation, the following practical points should be observed:

a. The boiling flask should not be more than half full.

b. Boiling stones are added to the liquid to prevent bumping.

c. Each ground joint should be greased to ensure a completely sealed system.

d. Cooling water in the condenser should enter at the lower end and exit at the upper end. This ensures that the condenser jacket is always full of water.

e. The bulb of the thermometer should be below the opening of the side arm so as to measure the temperature at which liquid and vapour are in equilibrium.

f. Heat sources used depend on the nature of the liquid. A water bath isused to distil low-boiling and flammable liquids, while a burner with wire gauze is used for higher boiling and less flammable liquids.

OBJECTIVES:

1. Distilling a pure liquid (acetone) and determining its boiling point.

2. Separating a mixture of acetone and water by simple distillation.

3. Separating a mixture of acetone and water by fractional distillation. Having done this experiment you should be able to compare the efficiency of each type of distillation.

EXPERIMENTAL

MATERIALS NEEDED	Glassware: Round-bottomed flasks (50 and 100 mL), distillation head, adapter, condenser, 2 Erlenmeyer flasks (50 mL), fractionating column, 2 stands, wire gauze, 2 clamps, 2 clamp holders, large beaker for water bath, graduated cylinder (10 mL). Chemicals: 30 mL acetone, boiling stones, grease

DETERMINATION OF BOILING POINT OF PURE ACETONE:

Arrange a simple distillation apparatus as shown in Figure 11 using a water bath as a heat source. Introduce about 20 mL of a liquid (e.g. acetone) and a few boiling stones in a 50 mL round-bottomed flask. Heat gently so that the distillate collects in the receiver drop by drop. Make sure that there is a drop of liquid hanging from the bulb of the thermometer to ensure that the thermometer is reading the correct boiling point. Absence of this drop indicates superheating. Wait until 1-2 mL of the distillate have been collected before recording the temperature. Continue the distillation until about 2 mL of residue are left in the distillation flask, and record the temperature again. Keep the acetone for the following part.

SEPARATION OF A MIXTURE:

1. Simple Distillation: Make a mixture of two liquids (e.g. acetone-water) 20 mL each and pour it into a 100 mL round-bottomed flask. Carry out a simple distillation as before and collect five fractions in the following boiling ranges: 50-62, 62-72, 72-82, 82-95 plus the fifthfraction which is the residue. Measure the volume of each fraction and record the results in the report sheet.

2. Fractional Distillation. Combine the five fractions and pour into a100 mL round-bottomed flask, attach the fractionating column and proceed as for simple distillation. Measure the volume of each fraction as before and record your results.

QUESTIONS:

1) A pure liquid has a constant boiling point, but a liquid with a constant boiling point is not necessarily pure. Explain.
2) What is the effect of each of the following on the observed bp?

   a)The thermometer is not kept moist with condensate.

   b)   The presence of a non-volatile impurity.
3) What effect does a reduction of the external pressure have on the boiling point?
4) Why is it important that cooling water enters at the lower end and exits at the upper end of the condenser jacket, and not *vice versa*?
5) During a distillation why the distilling flask should be filled to two thirds of its capacity only?

QUALITATIVE ELEMENTAL ANALYSIS

The identification of organic substances is a major subject in organic chemistry. It involves physical (melting point, boiling point, solubility....etc.) as well as chemical investigation.

This experiment is concerned with qualitative elemental analysis of organic compounds. A knowledge of the elemental composition of a substance is useful in planning the identification procedure. For example, a compound which is found to contain no nitrogen can certainly not be an amine, a nitro compound, a nitrile ...etc. and one would therefore, in such a case, direct his search towards other functional groups.

IGNITION TEST:

Most organic substances burn with characteristic flames that help in determining their elemental constituents:

1. Aromatic and unsaturated hydrocarbons, which have relatively high carbon content, burn with yellow, sooty flames.
2. Saturated hydrocarbons burn with flames that are yellow but much less sooty.
3. Oxygen-containing compounds (e.g. alcohols, ethers, ..etc.) burn with bluish flames. The blue colour is however not pronounced if the oxygen/carbon ratio is low.
4. Metal-containing compounds, when ignited, leave residues (ash) due to the formation of non-volatile metal oxides.
5. A high content of halogen (e.g. chloroform, carbon

tetrachloride) or oxygen (e.g. sugars) drastically reduces the flammability of the substance.

TEST FOR SULFUR, NITROGEN, AND HALOGEN:

The detection of these elements, is based on simple chemical tests, provided that they are first converted to their ionic forms. This is achieved by heatingthe organic substance with sodium metal which converts the above elements to the corresponding sodium salts: sulphur is converted to sodium sulphide, nitrogen to sodium cyanide and halogen to sodium halide.

$$\text{C, H, S, N, X (covalent forms)} + \text{Na} \longrightarrow \text{NaCN, Na}_2\text{S, NaX (ionic)}$$

Sulphur

Sulphide anion is detected by lead acetate. A black precipitate of lead sulphide (PbS) is formed if S is present. The solution should be made acidic prior to the addition of lead acetate so as to eliminate inter-fearing hydroxide ions.

$$Pb^{+2} + S^{-2} \longrightarrow PbS \text{(black ppt)}$$

Nitrogen:

Sodium cyanide reacts with ferrous ammonium sulphate to yield sodium Ferro cyanide, which reacts with ferric salts to give a deep blue colour due to the presence of sodium ferriferrocyanide (Prussian blue). A faint green colour indicates incomplete fusion of the organic substance with sodium.

$$6NaCN + Fe(NH_4)_2(SO_4)_2 \longrightarrow Na_4Fe(CN)_6 + Na_2SO_4 + (NH_4)_2SO_4$$

sodium   ferrous ammonium   sodium
Cyanide   sulfate   ferrocyanide

$$Na_4Fe(CN)_6 + Fe^{+3} \longrightarrow NaFe[Fe(CN)_6] + 3Na^+$$

sodiumferriferrocyanide
(Prussian blue)

Halogens:

Sodium fusion converts halogens to sodium halides, which are easily detected by silver nitrate. Chloride produces a white precipitate which dissolves in concentrated ammonia, bromide gives a pale yellow precipitatewhich is partially soluble in ammonia, and iodide gives a yellow precipitate which is insoluble in ammonia.

Hydroxide, sulphide, and cyanide ions interfere in this test since they form insoluble silver salts. It is therefore essential to get rid of such ions before the addition of silver nitrate. Hydroxide ions are neutralized by acidification with nitric acid while sulphide and cyanide ions are removed by boiling the acidified solution for a few minutes to expel the gaseous hydrogen sulphide (H_2S) and hydrogen cyanide (HCN). In practice, the sulphur and nitrogen tests are made before the halogen test. In case either of them is present, the solution should be boiled after acidification and before adding the silver nitrate.

28

EXPERIMENTAL

MATERIALS NEEDED	**Glassware:**Spatula, ignition test tube, Test tubes
	Chemicals: Sodium, ethyl alcohol, distilled water, charcoal, lead acetate solution, 5% ferrous sulfate solution, 10% potassium fluoride solution, 5% ferric chloride solution, dilute sulphuric acid, litmus paper, dilute nitric acid, silver nitrate,
	concentrated ammonia.

Ignition Test:

Perform the ignition test on some selected organic substances by holding a spatula containing a small amount of the sample (50 mg of a solid or 2-3 drops of a liquid) over a small flame. When the substance catches fire, remove the spatula out of the flame, and carefully observe the characteristics of the flame. Also note if the substance leaves a residue (ash) after burning.

Sodium Fusion:

Note: Sodium is a dangerous chemical that should never be allowed to come into contact with water or skin. Use forceps to handle. Excess sodium is destroyed with ethyl alcohol.

Place a small amount of the substance (0.1 g) and a small piece of sodium into a small test tube (ignition tube). Heat the lower part of the tube over a hot flame (use test tube holder) until no more fumes are formed. Continue heating until thelower part of the tube is red hot. Quickly drop the tube into a small beaker containing 20 mL of distilled water. Use a glass rod to stir the contents of the broken tube to assists dissolution of any salts. Boil the mixture for 1-2 minutes then remove the insoluble materials by filtration. If the filtrate is coloured, decolourize with charcoal and filter to remove the charcoal. Save this "stock solution" for the following tests.

Sulphur Test:

Acidify 3 mL of the stock solution with dilute acetic acid until acidic to litmus. Add 5 drops of lead acetate solution, shake and observe the result.

Nitrogen Test:

To 3 mL of the stock solution, add 5 drops of 5% ferrous sulphate and 5 drops of 10% potassium fluoride. Boil gently for 5 seconds, and allow the tube to cool, then add 3 drops of 5% ferric chloride. Finally, add sufficient dilute sulphuric acid to dissolve the insoluble iron hydroxides and make the solution acidic to litmus. Observe the result.

Halogen Test:

Acidify 3 mL of the stock solution with dilute nitric acid. If sulphur or nitrogen, or both, are present, boil for 4-5 minutes. Add 5 drops of silver nitrate and observe the result. Note the colour of the precipitate formed. To determine the nature of the halide present, boil the contents of the tube for a short time so that the precipitate coagulates. Decant the supernatant liquid and wash the residue with 2 mL of distilled water. Test the solubility of the precipitate in concentrated ammonia (2 mL).

RECRYSTALLIZATION

A Purification Technique for Solids

INTRODUCTION:

Crystallization may be defined as the process in which a solid compound precipitates from a saturated solution in the form of crystals. Saturation is usually effected through cooling or evaporation. In certain cases, recrystallization may be used for the separation of a solid mixture (fractional recrystallization).

Purification by recrystallization depends on the following facts:

a. Different solids have different solubility in a given solvent.

b. Most solids are more soluble in hot than in cold solvents.

When the impure solid is dissolved in a minimum volume of a suitable hot solvents and the resulting solution is gradually cooled, saturation and eventual crystallization of the pure compound occurs.

Impurities in a solid are of twokinds: soluble and insoluble and recrystallization involves the removal of both to purify a solid. Insoluble impurities are first removed by gravity filtration of the hot solution while the soluble impurities remain dissolved in the cold saturated solution (*mother liquor*) after precipitation of the desired compound. The pure crystals are separated from the supernatant liquid by suction filtration. After drying, the purity is checked by a melting point determination.

Crystal formation is a selective process and only molecules of the same substance can fit into the crystal lattice, excluding foreign molecules (impurities) which remain in solution (Figure 13).

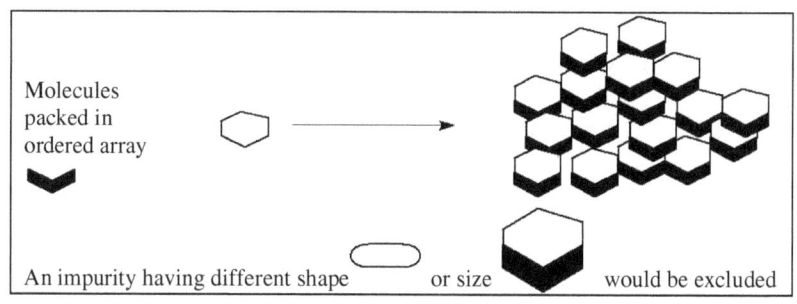

Figure 13. The crystallization process.

The solubility of a solid solute in a solvent is determined by two factors:

a. The relative polarities of the solvent and solute. *"Like dissolves like"* isthe best summary of solubility behaviour. Polar solvents dissolve polar solutes and non-polar solvents dissolve non-polar solutes. For example, solutes that contain polar groups like OH, NH_2, and COOH dissolve in polar solvents like water, methanol, and ethanol while hydrocarbons and their halogenated derivatives are non-polar and dissolve in non-polar solvents like chloroform, carbon tetrachloride, hexane and petroleum ether.

b. The lattice energy of the crystalline solute. The crystal lattice, holdingsolute molecules together in the solid state, is broken down upon dissolution. The necessary energy is provided through *"solvation"* of the solute by solvent molecules. The stability of a crystal lattice is roughly reflected by the melting point: a high melting point indicates a high lattice energy, and *vice versa*. For a given set of isomers, the higher the melting point, the less soluble the substance is in a given solvent.

A suitable solvent for recrystallization should possess

the following important properties:

    a. Dissolve a large amount of the solid to be purified at hightemperatures, but very little at room temperature.

    b. Dissolve impurities readily at low temperatures or not at all even atthe boiling point.

    *c.*Not react with the substance to be purified.

    d. Evaporate readily from the crystals, i.e., be relatively volatile.

If two or more solvents appear to be equally suitable, it is preferable to choose a solvent which is non-flammable, non-toxic, and cheap.

GENERALIZED EXPERIMENTAL PROCEDURE:

Recrystallization involves the following sequence of steps:

1. Selection of a suitable solvent.
2. Preparation of the hot solution and "*decolorization*" if necessary.
3. Filtration of the hot solution to remove insoluble impurities (and charcoal).
4. Cooling to effect crystallization.
5. Collection (cold filtration), washing, and drying of the crystals.

Each step will now be discussed more fully.

1. Selection of the Solvent: The suitable solvent is determinedexperimentally through solubility tests.

This is done by shaking about 0.1 g of the powdered solid with 2 mL of the given solvent in a dry test tube. If all the solid has nearly dissolved in the cold solvent, the solvent is considered unsuitable.

If not, the mixture is heated gently to the boiling point with stirring. (water bath for flammable solvents). If most of the solid did not dissolve, the solvent is also unsuitable.

If a substance is found to be too soluble in one solvent and insoluble in another, then a mixture of both solvents (solvent pair) may be used. In such cases the two solvents must be completely miscible. The compound to be recrystallized is first dissolved in the solvent in which it is very soluble, then the other solvent is added gradually, with heating, until a slight turbidity occurs. The solution is then allowed to stand at room temperature to effect slow crystallization before chilling in ice.

Table 2. Common solvents for recrystallization.

Solvent	b.p	Particulars of Solvent
Water	100	to be used whenever suitable
Methanol	65	flammable; toxic
Ethanol	78	flammable
Acetone	56	flammable
Ethyl acetate	78	flammable
Chloroform	61	non-flammable; vapor toxic
Benzene	80	flammable; vapor highly toxic
Cyclohexane	81	flammable

2. **Preparation of the Solution:** To prepare the hot solution, the solid isplaced in an Erlenmeyer flask and the selected solvent is added in small portions. The mixture is stirred and heated to boiling after each addition, until the solid dissolves completely. A slight excess of the solvent is usually added to compensate for any losses (through evaporation) during filtration.

Decolorizing charcoal may be added at this stage if

the solution is coloured due to coloured impurities. The flask should be removed from the heat source before adding charcoal to it, otherwise bumping will occur.

3. Hot Filtration (Gravity Filtration): Filtration of the hot solution isnecessary to remove insoluble impurities. A fluted filter paper and a short-stem funnel (Figure 15) allow rapid filtration and avoid premature crystallization inside the stem and on the filter paper.

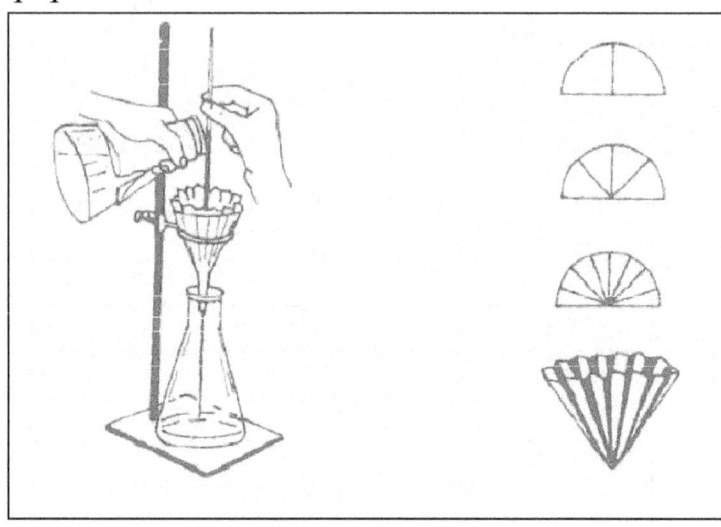

Figure 15.Rapid filtration of a hot solution using a fluted filter paper.

4. **Cooling**: To induce crystallization, the clear, hot filtrate is allowed tocool down to room temperature, undisturbed, until a large amount of crystals has formed. The mixture is finally chilled in ice to complete crystallization.

5. **Collecting and Drying of Crystals:**The crystals are collected by suctionfiltration (cold filtration) using a Buchner funnel to ensure rapid and complete removal of the solvent. The crystals are then washed with a few milliliters of fresh, ice-cold solvent to get rid of the last traces of mother liquor. The crystals

are finally dried in an oven or allowed to air-dry, in case the melting point is low, by spreading them over a sheet of paper.

OBJECTIVES:

1. Selection of suitable solvents for recrystallization.
2. Recrystallization of an unknown compound.

EXPERIMENTAL

MATERIALS NEEDED	Glassware: 5 test tubes, 2 Erlenmeyer flasks (100 mL), filter flask, filter funnel, Buchner funnel, stand, wire gauze, clamp, clamp holder, water bath, melting point apparatus. Chemicals: 30 mL of each of ligroin, ethanol, 0.6 g of each of salicylic acid, anthracene, sodium benzoate,1.0 g of unknown (Table 1).

SELECTION OF SOLVENT:

Perform solubility tests on anthracene, salicylic acid, and sodium benzoate in water, alcohol, and ligroin as follows:

With a spatula take about 0.1 g of the powdered solid and place in a dry test tube. Start by dissolving it in 2 mL of solvent with stirring. If insoluble, heat the mixture to boiling (water bath for flammable solvents) and observe the solubility. The results should allow the selection of a suitable solvent for each compound.

RECRYSTALLIZATION OF AN UNKNOWN:

Obtain an unknown and perform solubility tests as described above to choose the best crystallizing solvent. Recrystallize 1.0 g of this unknown from the solvent you have selected. Make sure you useonly the minimum volume of solvent, otherwise the amount of recovered product will be small. Determine the weight and melting point of the purified unknown compound.

QUESTIONS:

1) Why is it preferable to allow the hot solution to cool gradually instead of chilling immediately in an ice-bath?
2) Mention three properties a solvent should have to be suitable for recrystallizing a particular organic compound?
3) For what purpose is charcoal used in recrystallization?
4) How are insoluble impurities removed during recrystallization?
5) Why must the flask and funnel be heated before the hot solution is filtered?
6) Why is it important to minimize evaporation during the filtration of the hot solution?

EXTRACTION

A Separation and Isolation Technique

INTRODUCTION:

Extraction is the separation of a substance from a mixture by means of a solvent that preferentially dissolves that substance. If the substance is extracted from a solid phase, the process is called solid-liquid extraction, as in the isolation of caffeine from tea leaves by means of hot water. Extraction of a substance from a liquid phase is called liquid-liquid extraction. The most common applications of this latter technique are:

a. The recovery of an organic product from a reaction mixture containing excess unreacted materials and by-products.

b. Isolation of an organic substance from its natural source, such as a plant.

Liquid-Liquid Extraction: This is the most common type of extraction. Itinvolves shaking the liquid mixture with an immiscible solvent which preferentially dissolves the desired compound. On standing, the two immiscible phases (usually organic and aqueous) form two separate layers (upper and lower) that can be separated by means of a separatory funnel. The various solutes in the mixture distribute themselves between the organic and aqueous phases according to their relative solubilities in each solvent. At equilibrium, the ratio of the concentration (C) or solubility (S) of the substance in the organic phase, (C_o or S_o) to that in the aqueous phase (C_w or S_w) is called the distribution coefficient (K_D).

$$K_D = \frac{C_o}{C_w} = \frac{S_o}{S_w}$$

A large distribution coefficient implies that the compound is much more soluble in the organic phase than in the aqueous one and, in this case, a single extraction suffices to remove the desired compound from the mixture. When K_D is small, it means that the compound distributes itselfmore evenly in both phases, so that repeated extractions are required to recover such a compound from the aqueous mixture.

In general, it is more efficient to divide the total volume of extracting solvent over several extractions than to use the whole volume in a single extraction.

GENERAL EXPERIMENTAL CONSIDERATIONS:

Choice of Solvent: A solvent used for extraction should have the followingcharacteristics:

1. Immiscible with the liquid in which the solute is present.
2. Readily dissolve the solute to be extracted.
3. Extract little or none of the impurities and other compounds presentin the mixture.
4. Non-flammable, non-toxic, cheap and easily removable from the solute after extraction (*i.e.*, volatile).

Salting-out: Extraction of organic compounds from aqueous mixtures isusually improved by saturating the aqueous phase with a salt such as NaCl or Na_2CO_3. This phenomenon is called salting-out and has the followingeffects:

1. Decreases the solubility of organic compounds in the saturated aqueous phase.
2. Decreases the solubility of the organic and aqueous phases in each other, thus improving their separation. This is particularly useful in breaking up emulsions.

Emulsions: In certain cases, the two immiscible phases do not separatecleanly into two distinct layers; instead, they form an emulsion which, once formed, is usually difficult to break. It is therefore advisable to prevent the formation of emulsions during extraction. This is best achieved by avoiding vigorous shaking of the layers whenever an emulsion is expected to form

(*e.g.*, when alkaline aqueous solutions are extracted with chloroform or dichloromethane). If an emulsion still forms one can often break it by:

1. Stirring the emulsified layer gently with a glass rod.
2. Saturating the aqueous layer with a salt.
3. Centrifugation.

Drying Agents: The organic phase often shows turbidity due to thepresence of traces of water from the aqueous phase. Anhydrous $CaCl_2$, $MgSO_4$, or Na_2SO_4 may be used as drying agents which absorb the traces of water present in the organic phase. When dry, the organic phase becomes clear.

Acid-Base Extraction:

Mixtures of organic acids and bases are commonly separated by acid-base extraction. Such compounds are converted to their salts by treatment with acid or base. Unlike the original compounds, the corresponding salts are usually soluble in water, thus enabling their transfer from the organic phase to the aqueous layer. After separation of the layers, the organic acid or base is recovered by neutralization of the aqueous layer. Since the acid or base is insoluble in water it precipitates out, and is collected by filtration. A flow diagram for such a separation is shown in Figure 16.

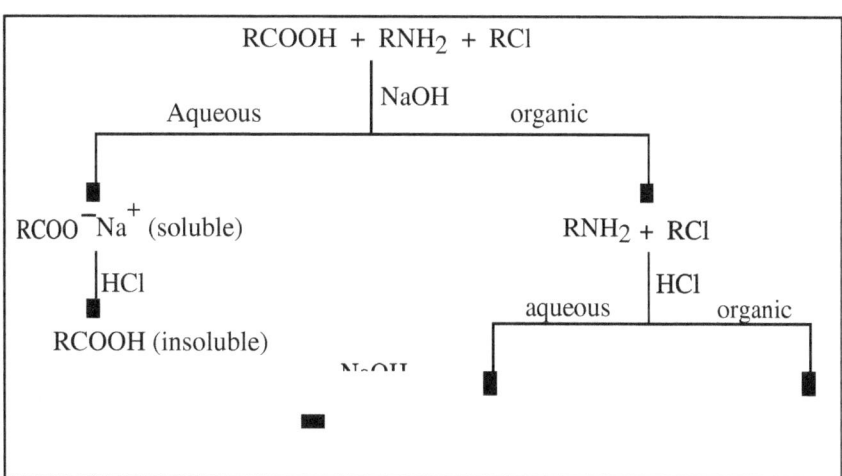

$$RNH_2 \xrightarrow{\hspace{2cm}} RNH_3^+ \overset{-}{Cl} \qquad RCl$$
(insoluble) \qquad\qquad (soluble)

Figure 16.Flow chart for separation of a mixture by acid base extraction.

APPARATUS AND PROCEDURE:

The set-up for an extraction using the separatory funnel is shown in Figure 17. The stopcock and stopper should be greased before use. The separatory funnel is held upside down in both hands, such that the stopper is firmly held in the left hand, while the right hand controls the stopcock (Figure 17). The funnel is shaken gentlyat the beginningand vented periodically through the stopcock to release excess pressure which builds up inside. Once there is no more pressure build-up in the funnel, shaking may be more vigorous to ensure good extraction.The mixture is left standing for sometime until the two layers are well separated. The lower layer is drained through the stopcock, while the upper layer is poured through the top of the funnel into a separate container.

Figure 17. Correct position for holding a separatory funnel.

To determine whether a given layer is organic or aqueous,

place a few drops of it on a watch glass containing a few milliliters of water and check the solubility. As a precaution, never discard any layer before you get your product.

OBJECTIVES:

1. Isolating caffeine from tea leaves.
2. Separating a two-component mixture (acid and neutral).

EXPERIMENTAL

MATERIALS NEEDED	Glassware: Beaker (600 mL), 2 beakers (100 mL), 2 Erlenmeyer flasks (100 mL), stand, wire gauze, clamp, ring, clamp holder, graduated cylinder (10 mL), separatory funnel (100 mL), cheesecloth 20x20 cm. Chemicals: 15 g Tea leaves, 10 g sodium carbonate, 30 mL dichloromethane, 1.0 g anhydrous sodium sulfate, 1 g benzoic acid, 1 g p-dichlorobenzene, 25 mL ether, 40 mL of 10% sodium hydroxide, 20 mL concentrated hydro chloric acid, anhydrous calcium chloride, blue litmus paper.

Caffeine is an organic compound present in the fruit and bark of some plants, as well as in tea leaves, coffee, cocoa and cola beans. The caffeine content in dried tea leaves is about 3-4%

Caffeine belongs to a family of basic, nitrogen-containing, cyclic compounds called alkaloids. It is a mild stimulant and is used as such in many drugs and analgesics. The solubility of caffeine at room temperature is 2.2 g/100 mL of water and 18 g/100 mL of chloroform.

In this experiment, you will extract caffeine from tea leaves with hot water. This treatment also extracts tannins, a class of acidic organic compounds, also present in the leaves. Sodium carbonate is used to remove the acidic tannins by converting them to water-soluble salts.

Procedure: You will be provided with a large tea bag containingabout 15 g dry tea leaves. Place the tea bag in a 600 mL beaker, add 10 g of sodium carbonate and 150 mL of water, and boil the mixturegently for 20 minutes. Cool the dark brown aqueous solution to room temperature and squeeze the tea bag to extract the liquid fully before discarding the bag.

Transfer the dark solution to a separatory funnel and extract twice with 15 mL portions of dichloromethane. Avoid vigorous shaking of the funnel since emulsions may form readily; instead, swirl the funnel gently or turn it upside down several times. After each extraction drain the denser dichloromethane layer into a small flask. Dry the combined organic extracts with anhydrous sodium sulphate until the solution is clear. Decant the dichloromethane into a small beaker and evaporate to dryness over a water bath in the fume hood. Do not heat the residue any longer than necessary since caffeine decomposes readily. Weigh the crude caffeine and calculate its percentage in the tea leaves. Determine the melting point of your product.

SEPARATION OF A TWO COMPONENT MIXTURE:
In this part, a mixture of benzoic acid and *p*-dichlorobenzene (or any other neutral compound such as naphthalene, *p*-dimethoxybenzene or diethoxybenzene) will be separated into its components by means of extraction.

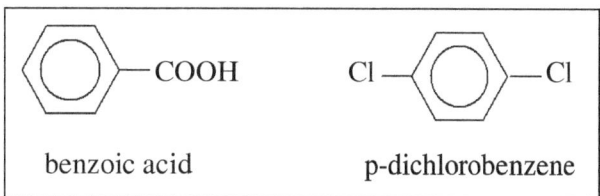

benzoic acid   p-dichlorobenzene

Procedure:Obtain from your instructor a 2.0 g sample of the twocomponent mixture, dissolve it in 25 mL of ether, and pour the solution into a 100 mL separatory funnel. To extract the benzoic acid from the mixture shake the ether solution with 20 mL of 10% NaOH solution.

Draw off the lower (aqueous) layer into a flask and repeat extraction of the ether solution with another 20 mL of 10% NaOH solution followed by 10 mL of water. Combine the three aqueous extracts and cool in an ice bath. Neutralize the cold aqueous phase by adding concentrated HCl until the solution is acid to blue litmus paper (about 8 mL). Collect the precipitated benzoic acid by suction filtration. Dry the product, determine its weight and its melting point. Pour the remaining ether solution (which contains the neutral component) into a small flask and add enough anhydrous calcium chloride to remove any traces of water. Decant the dry ether solution into a small weighed beaker and evaporate the solvent in the fume hood (use a low temperature water bath since *p*-dichlorobenzene may sublime). Determine the weight of your product.

STEAM DISTILLATION

A method for Purification and Separation of Liquids and Solids

INTRODUCTION:

Steam distillation is the distillation of slightly volatile, water-insoluble substances by means of steam.

For a mixture of two completely immiscible compounds A and B, the total vapour pressure (P_T) can be expressed, as usual, in terms of partial pressures p_A and p_B:

$$P_T = p_A + p_B = P_A^o . X_A + P_B^o . X_B$$

X_A and X_B are the mole fractions of A and B in the liquid mixture. However, since A and B are immiscible, each behaves independently of the other and therefore $X_A = X_B = 1$. The above expression thus becomes:

$$P_T = p_A \quad p_B = P_A^o + P_B^o = constant$$

This means that the total vapour pressure of such a heterogeneous (two-phase) mixture is constant and depends only on the vapour pressure of pure A and pure B at a given temperature. These vapour pressures are completely independent of the relative amounts of A and B in the mixture. Unless either $P_A^°$ or $P_B^°$ is zero, the vapour pressure of the mixture (at a given temperature) is higher than the vapour pressure of either component alone. The boiling point of the mixture will be the temperature at which the total vapour pressure equals atmospheric pressure and it is obviously lower than the boiling point of either pure A or pure B. Refer to Figure 18 which shows the vapour pressure-temperature relationship for an immiscible mixture of bromobenzene and water.

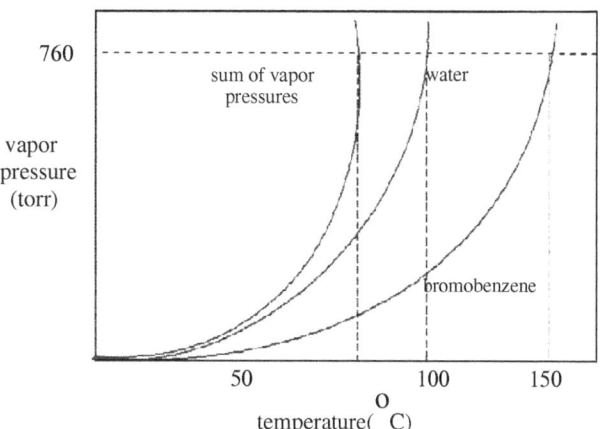

Figure 18. Vapour pressure-temperature curves for two immiscible liquids.

Steam distillation always takes place at a temperature below the boiling point of water. This makes it possible to distil high-boiling, steam-volatile, organic compounds at temperatures below $100^{\circ}C$, thus avoiding possible decomposition of such compounds. At equilibrium, the composition of the vapour above the liquid mixture can be expressed as:

$$\frac{\text{moles of A}}{\text{moles of B}}_{\text{(in vapor)}} = \frac{P_A}{p_B} = \frac{P_A^{\ 0} \cdot X_A}{P_B^{\ 0} \cdot X_B} = \frac{P_A^{\ 0}}{P_B^{\ 0}} \quad (\text{remember } X_A = X_B = 1)$$

The weight ratio of A and B in the vapor will depend not only on their molar ratios but also on their molar masses. Since the molar mass of the organic substance is usually much higher than that of water, significant amounts of the organic substance will distil over, even though its vapor pressure may be relatively low.

$$\frac{\text{Weight of A}}{\text{Weight of B}} = \frac{P_A^{\ 0} \cdot (MW)_A}{P_B^{\ 0} \cdot (MW)_B}$$

Steam distillation is convenient for the purification of high-boiling compounds by low-temperature distillation and so replaces vacuum distillation. An inherent disadvantage of this

46

technique is that it is limited to substances with the following properties:

1. Steam-volatile, *i.e.*, those that have appreciable vapour pressures at the temperature of steam distillation (between 90 and 95°C).
2. Immiscible with water.
3. Inert toward steam and stable under the conditions of steam distillation.

Steam distillation finds application in the separation of such compounds from mixtures containing non-volatile impurities, as well as in the isolation of steam-volatile organic compounds from natural sources (plants). This technique is not restricted to liquids but can be applied to solids as well, provided they fulfil the above mentioned conditions.

ESSENTIAL OILS:

The characteristic aromas of plants are due to volatile essential oils, which are used as a source of fragrances and flavourings. These oils occur in all parts of the plant. Essential oils are generally complex mixtures of hydrocarbons, alcohols, and carbonyl compounds, mostly belonging to the broad group of plant products known as terpenes. Essential oils are best isolated from the plant tissue by steam distillation. In this experiment, the essential oil of one of four widely used spices (anise, caraway, cumin and cloves) will be isolated by steam distillation.

Anise Oil: The essential oil obtained from anise, a popular flavouring forcookies, is predominantly *trans*-1-(*p*-methoxyphenyl)propene (*anethole*) which comprises 80-90% of the oil. A minor component is the double bond isomer, 3-(*p*-methoxyphenyl)propene (*p-allylanisole*). Anethole has a melting point near room temperature, and the oil crystallizes on chilling.

Caraway Oil: The essential oil of caraway, comprising 1-3% by weight of the dry seeds, contains two principal compounds which are carvone and limonene.

Cumin Oil: The major volatile constituent of cumin is *p*-isopropylbenzaldehyde (*cumin aldehyde*). Cumin also contains limonene.

Clove Oil: Oil of cloves is rich in 4-allyl-2-methoxyphenol (*eugenol*) whichis used medicinally as a dental antiseptic and analgesic.

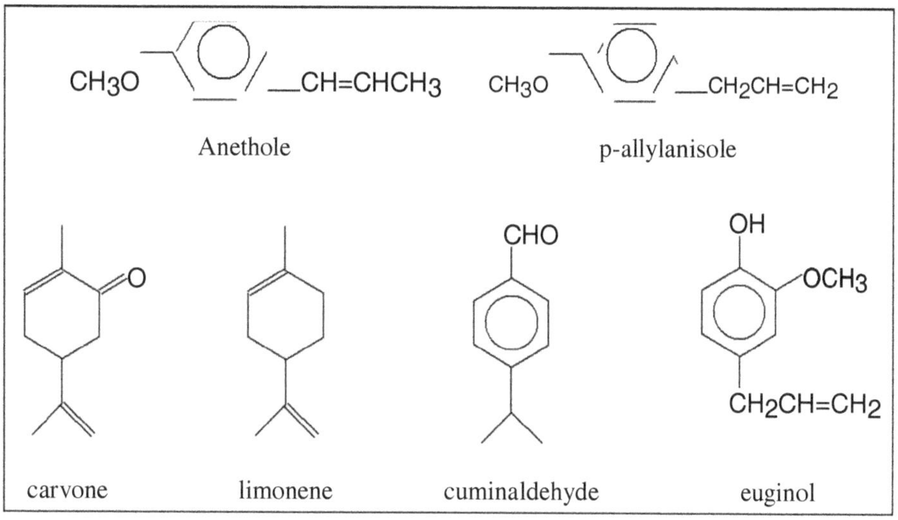

APPARATUS AND PROCEDURE:

A typical apparatus for steam distillation is shown in Figure 19. The apparatus is basically the same as that used for simple distillation. Here, a large volume of water is introduced into the distillation flask to provide the steam. A strong flame is used to heat the contents of the flask, which should be large enough to enable vigorous boiling of the mixture. If the substance being distilled is a solid, it often solidifies and accumulates inside the condenser. In such cases use an air condenser or stop the flow of cooling water until the solid melts and eventually collects in the receiver.

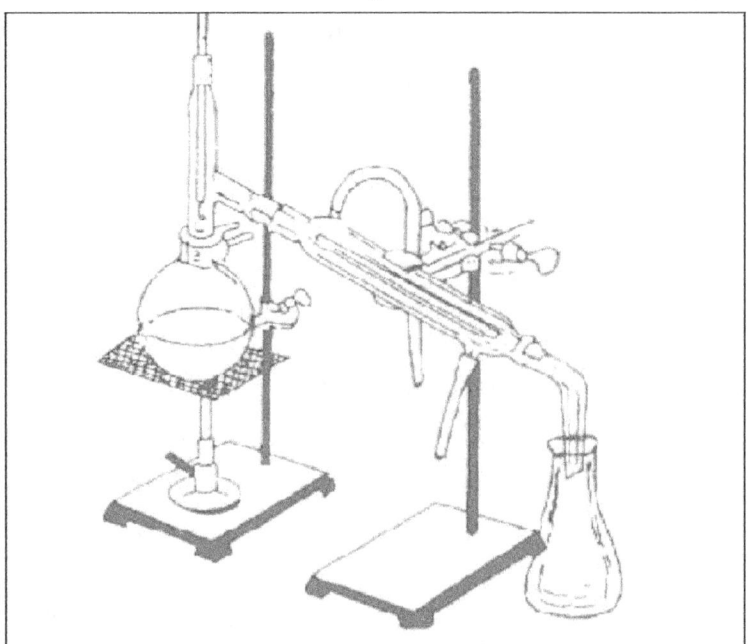

Figure 19.Steam distillation set-up.

OBJECTIVES:

1. Steam distillation of bromobenzene
2. Isolation of essential oils from cloves or the seeds of one of the following: anise, caraway, or cumin.

Having done this experiment you will have examined the theoretical features of steam distillation, and seen the application of steam distillation to the isolation of essential oils from plants.

EXPERIMENTAL

MATERIALS NEEDED	Glassware: Round-bottomed flasks (250 mL, 500 mL), distillation head, adapter, condenser, Erlenmeyer flask (250 mL), beaker (100 mL), 2 stands, wire gauze, 2 clamps, 2 clamp holders, graduated cylinder (10 mL), separatory funnel (250 mL), water bath. Chemicals: 25 mL Bromobenzene, 15 g sodium chloride, 1g anhydrous sodium sulfate, 50 mL dichloromethane and 25 g of one of the following: anise seeds, caraway seeds, cumin seeds, orcloves.

STEAM DISTILLATION OF BROMOBENZENE:

This part is intended to demonstrate the theoretical features of steam distillation using bromobenzene as an example. Introduce 25 mL of bromobenzene and about 100 mL of water into a 250 mL round-bottomed flask, connect a condenser, and start the distillation in the usual manner. Discard the first 5 mL of distillate then collect three successive 10 mL fractions. Transfer the fractions, one at a time, to a measuring cylinder and wait until the layers are clearly separated before recording the volume of each layer. Once you finish the experiment pour the fractions containing bromobenzene into the residue bottle.

ISOLATION OýF ESSENTIAL OILS:

Place 25g of the ground spice and 200 mL of water in a 500 mL round-bottomed flask. Connect the flask to a distillation apparatus and heat the mixture to boiling. Collect the distillate until no more droplets of oil come over; a minimum of 150 mL should be collected. Add 10-15 g of sodium chloride and stir until completely dissolved.

Cool the distillate to room temperature, transfer to a 250 mL separatory funnel, and extract with 15 mL of dichloromethane. Separate the layers and collect the organic layer in a small flask. Repeat the extraction twice with another 15 mL each of dichloromethane and combine the organic extracts. Dry the organic phase with anhydrous sodium sulphate until the solution is clear. Decant the dichloromethane solution into a small beaker. Evaporate the solvent on the steam bath, in the fume hood, until the solution has been concentrated to an oily residue. Weigh the oil and calculate the percentage yield based on the weight of the dried spice.

QUESTIONS:

1) Discuss the results of part 1, concerning the boiling point, and composition of the distillate. Compare these results with those obtained from the water-acetone mixture in the distillation experiment.

2) What are the advantages and disadvantages of steam distillation as a method of purification?

3) Suggest another possible method that might be used to obtain essential oils from the spices.

4) At 95.5°C, the vapour pressure of water is 641 mm, and that of bromobenzene is 119 mm. Calculate the molar ratio and the weight ratio of bromobenzene to water when a mixture of the two is distilled at 760 mm. Compare the answers with your experimental results. (Density of bromobenzene = 1.5 g/mL).

CHROMATOGRAPHY

A Separation and Purification Technique

INTRODUCTION:

Chromatography is a technique that may be used to separate the components of a mixture as well as to identify organic substances and examine their purity. Chromatography encompasses several techniques such as column, thin-layer, paper, gas liquid, etc. chromatography. Two principles are basically involved in chromatography: adsorption (as in thin-layer chromatography) and partition (as in paper chromatography), and certain terms are common to both types of chromatography.

In adsorption chromatography, separation depends on the selective desorption of the components of a mixture by the eluent (mobile phase) from the surface of a solid adsorbent (stationary phase). The adsorbent may be packed in a column (column chromatography) or spread as a thin layer on a glass plate as in thin-layer chromatography.

In partition chromatography, separation depends on partition of the components of a mixture between the stationary and mobile phases. The mobile phase may be a liquid (liquid-liquid partition chromatography) or a gas (gas-liquid partition chromatography).

ANALYSIS OF CHROMATOGRAMS:

In thin layer and paper chromatography, substances are characterized by their R_f-values (retardation factor). The R_f-value is a number (less than one) which is characteristic of a compound for a given adsorbent and developing solvent. It is defined as:

$$R_f = \frac{\text{distance traveled by the compound}}{\text{distance traveled by the solvent}}$$

In gas-liquid chromatography, compounds are characterized by their retention times.

THIN-LAYER CHROMATOGRAPHY (TLC):

This is one application of adsorption chromatography in which an adsorbent, usually silica gel or alumina, is spread out as a thin layer on an inert surface, such as a glass plate or microscope slide. The mixture is applied at one end of the coated plate and, as the mobile phase (a liquid) moves up the solid adsorbent by capillary action, the adsorbed components of the mixture get desorbed and carried along at different rates by the moving solvent.

Adsorption of the components of the mixture, on the surface of the adsorbent, occurs to differing extents depending on their structural features and polarity. The more strongly adsorbed a given compound is, the slower it is transported by the mobile phase, and conversely, the more weakly adsorbed the compound is, the faster it is transported up the stationary phase. The result is that the components of the mixture are separated into different zones or spots (Figure 20).

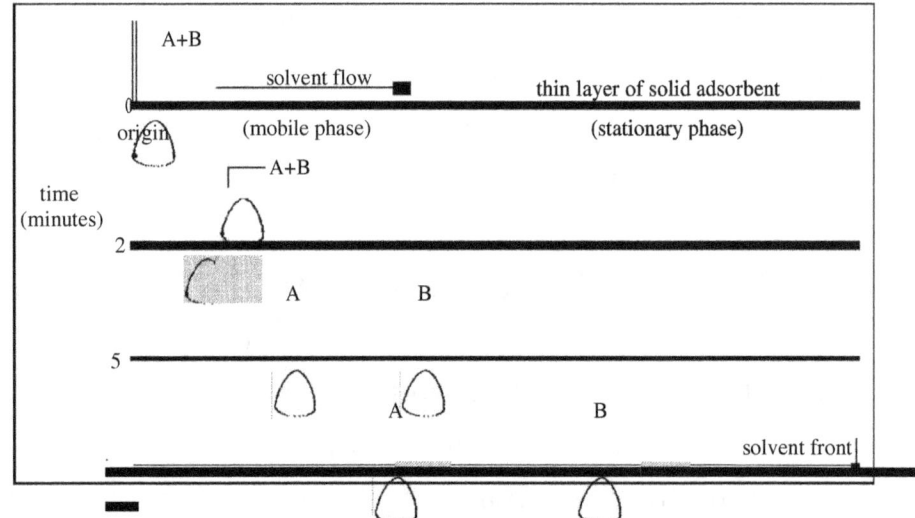

Figure 20. Separation by thin-layer chromatography

53

Separation by thin-layer chromatography depends on the kind and activity of the adsorbent (stationary phase), the polarity of the eluent (mobile phase) and on the chemical nature of the components of the mixture. The mostcommon adsorbents employed in *TLC* are silica ($SiO_2 \cdot xH_2O$) and alumina ($Al_2O_3 \cdot xH_2O$), and the activity of these adsorbents is largely determined by their water content. For a given adsorbent and compound, the greater the polarity of the eluent, the greater is its ability to dislodge a compound from the surface of the adsorbent, and therefore the higher the R_f-value.

Eluting power of solvents:
Acetic acid > Ethyl alcohol > Acetone > Diethyl ether > Dichloromethane > Hexane.

GENERALIZED EXPERIMENTAL PROCEDURE:

Preparation of *TLC* Plates:Large glass plates (20x20 cm) are commonlyused for quantitative separations, while microscope slides are usually used for qualitative purposes. Homogeneous slurry of the adsorbent in a volatile organic solvent (chloroform or dichloromethane) is poured over the glass plates and allowed to air-dry at room temperature. Microscope slides can be coated, two at a time, by dipping them into the slurry for some time then holding them vertically to air-dry. The jar of adsorbent must be shaken thoroughly before each use to homogenize the slurry.

Spotting: The mixture to be analysed is dissolved in a suitable solvent (1%solution). With a drawn capillary tube, a small amount of this solution is spotted on the *TLC* plate about 1 cm from the bottom (Figure 21). The spots should have a diameter not larger than 1-2 mm, since larger spots result in "tailing" and overlapping of close spots. Once the solvent evaporates from the spots, the plate is ready for developing.

Development of the Chromatogram: The eluent, also called developingsolvent, is chosen on the basis of the nature and polarity of the compounds being studied. It is best to choose the solvent that will give a satisfactory Separation within the range of 0.2-0.8 R_f values. The plate is placed in a developing chamber (*e.g.* a covered beaker) containing the solvent and lined with filter paper soaked in the solvent to help saturate the atmosphere with solvent vapours. When the solvent front reaches the finish line, the plate is removed from the beaker and placed on the bench top to air-dry.

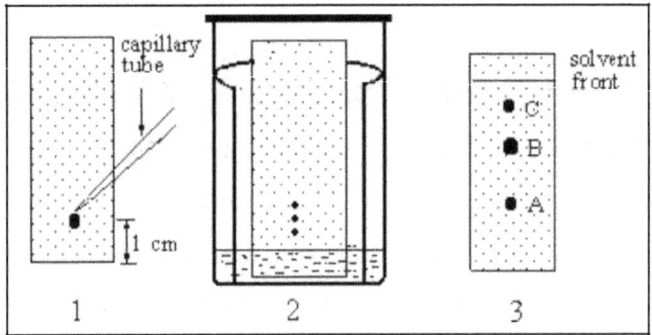

Figure 21.Steps in the TLC technique

Visualization of Spots: Compounds on the plate are located according totheir characteristics:

a. If the spots are coloured, they can be observed in ordinary light.
b. If the compounds are colourless, they can be seen under *UV*-light wherethey appear as dark spots on a white background.
c. Colourless spots may also be located with an indicator. Most organiccompounds form complexes with iodine

giving dark brown spots when the plate is exposed to iodine vapour. Sulphuric acid may also be used to make colourless spots visible. Most organic compounds turn black when sprayed with sulphuric acid.

OBJECTIVES:

1. Determining the R_f-value for *o*- and *p*-nitroaniline by *TLC*.
2. Separating a mixture of two dyes by paper chromatography.
3. Determining the constituents of an analgesic drug by *TLC*.

Having done this experiment you will have seen the use of *TLC* and paper chromatography in the separation of mixtures and in the characterization of organic compounds.

EXPERIMENTAL

MATERIALS NEEDED	<u>Glassware:</u> 2 Microscope slides,, capillary tubes, beaker (200 mL), Petri dish with cover, filter paper, UV lamp. <u>Chemicals:</u> stock solutions (1% in acetone) of *o*-nitrophenol, and *p*-nitrophenol, green dye, 5 mL dichloromethane, 20 mL isoprophyl alcohol, 1 mL methanol, 6 mL benzene, 3 mL ether, 1 mL acetic acid and analgesic tablet (Remin, Revanin, Paracetamol or Excedrin), stock solutions (5% in acetone) of aspirin, phenacetin, salicylamide, caffeine, acetaminophen.

TLC EXAMINATION OF ISOMERIC NITROANILINES:

Prepare the developing chamber by placing a filter paper inside a 200 mL beaker and adding 5 mL dichloromethane. Cover the beaker while preparing the *TLC* plates as described before. Dip a capillary tube into a 1% solution of *p*-nitroaniline in acetone and touch it tothe *TLC* plate at the origin. After the solvent has evaporated from the spot, place the slide in the developing chamber. When the solvent front has reached the finish line, remove the slide and allow the solvent to evaporate. Locate the center of the spot and calculate the R_f-value. Repeat the procedure with *o*-nitroaniline.

On a new slide, place the spots of *p*-nitroaniline and *o*-nitroaniline side by side at the origin so that the two compounds run parallel on the same slide. Calculate the R_f-value for each. Repeat using a mixture of the two compounds in addition to the single compounds as references. Place one spot of the mixture in the middle and a spot ofeach isomer on either side of the mixture. Note the resolution of the mixture into two spots, and compare the R_f-values obtained for the mixture and for the individual compounds.

ANALYSIS OF ANALGESIC DRUGS BY TLC:

In many non-prescription analgesic drugs such as Remin, Revacod, Revanin, Paracetamol, Excedrin etc.., the active ingredient is one or more of the compounds listed below. The constituents of an analgesic drug may be qualitatively determined by *TLC* using suitable reference samples. Introduce a small piece of an analgesic tablet into a test tube, add 1 mL of methanol and stir well. Allow the insolublematerial to settle and use the supernatant liquid for spotting on the *TLC* plate. Make dilute methanolic solutions of the referencecompounds and spot as many of them as possible on the same plate. Use a solvent mixture of benzene: ether: acetic acid (2:1:0.3) to develop the chromatograms. Examine the developed plates under ultraviolet light and determine the composition of the analgesic tablet.

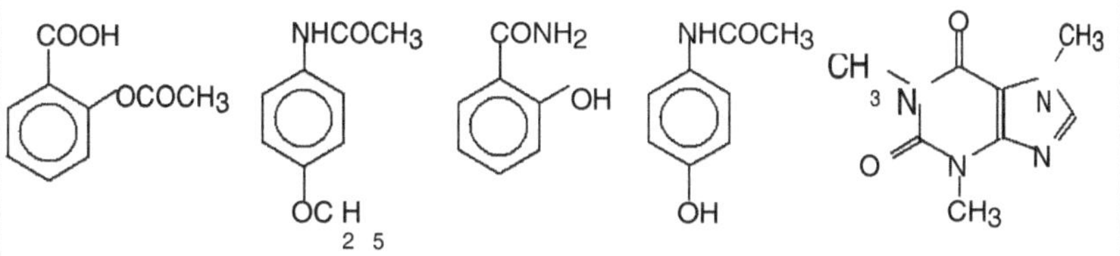

COOH
OCOCH3
Aspirin

NHCOCH3
OC_2H_5
phenacetin

CONH2
OH
Salicylamide

NHCOCH3
OH
acetoaminophen

CH_3
O
N
CH_3
N
N
O
CH3
caffeine

PAPER CHROMATOGRAPHIC ANALYSIS OF A DYE:

In this experiment green food colouring (composed of a yellow and a blue component) will be examined by paper chromatography using a Petri dish (10 cm in diameter) as a developing chamber.

Locate the center of a circular piece of filter paper by folding it in half. Make sure that the filter paper has a diameter slightly larger than that of the Petri dish. Using a melting point capillary tube withboth sides open, apply a small spot of food colouring at the center of the paper. Punch a small hole at the center and through it insert a small strip of filter paper rolled together to make a wick (Figure 22). Put 20-30 mL of the developing solvent (isopropyl alcohol - water 2:1) into the Petri disk and rest the filter paper on the rim of the dish making sure that the wick dips into the solvent. Cover the paper with the Petri dish cover and leave the chromatogram to develop undisturbed for 10 minutes until the colours separate into distinct circles. Remove the paper chromatogram and allow to air-dry. Calculate the R_f-values for the yellow and the blue dye.

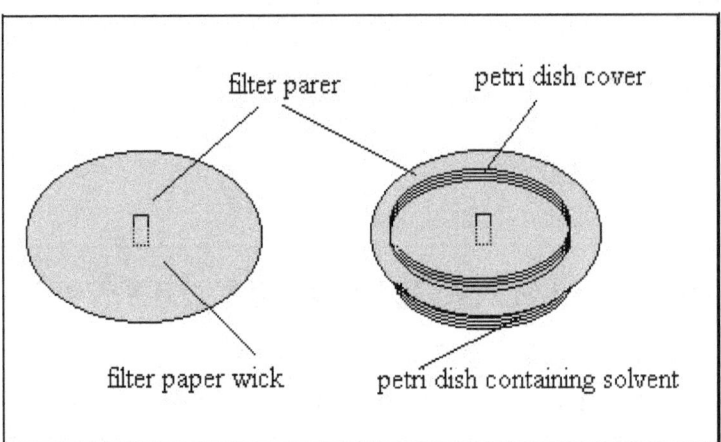

Figure 22. Paper chromatography using a Petri dish

QUESTIONS:

1) How will the following affect the *TLC* separation?

a) Too much sample applied.

b) Stationary phase is too active.

c) Forgetting to remove the plate when the solvent has reached the top of the plate.

d) Having too much solvent in the developing chamber so that its surface extends above the origin.

e) Polarity of the solvent being too high.

2) Which dye would be more soluble in *n*-propyl alcohol, the blue or the yellow? How can you tell?

3) Which compound is more strongly adsorbed on silica, *ortho*- or *para*-nitroaniline? Correlate the R_f values with the structures.

ALCOHOLS AND PHENOLS

Classification and Tests

I. ALCOHOLS:

Alcohols are classified as primary, secondary and tertiary according to the number of alkyl groups directly attached to the carbinol carbon.

$$
\begin{array}{ccc}
\underset{\substack{|\\ \text{OH}\\ \text{primary}}}{\overset{\substack{\text{H}\\ |}}{R-C-H}}
&
\underset{\substack{|\\ \text{OH}\\ \text{secondary}}}{\overset{\substack{\text{H}\\ |}}{R-C-R'}}
&
\underset{\substack{|\\ \text{OH}\\ \text{tertiary}}}{\overset{\substack{R''\\ |}}{R-C-R'}}
\end{array}
$$

Reactions of alcohols involve the breaking of either of two bonds: the O-H bond as in reactions with bases and esterification reactions, or the C-OH bond leading to dehydration and substitution reactions. In breaking the C-OH bond, protonation of the -OH group is essential to convert it from a poor leaving group to a better one.

Some physical and chemical properties of alcohols are examined in the following tests.

EXPERIMENTAL

MATERIALS NEEDED	Glassware: 4 Test tubes.
	Chemicals: 1 mL each of: ethanol,1-butanol, 2-butanol, 2-methyl-2-propanol, ethylene glycol,, sodium metal, phenolphthalein indicator, 15 mL potassium dichromate (1%), 0.5 mL sulfuric acid, 6.0 mL Lucas reagent, 15 mL iodoform reagent, 6 mL NaOH (10%),

1. Solubility in Water:

Alcohols of low molecular weight are water soluble due to their ability to form hydrogen bonds with water. Solubility in water decreases withincreasing molar mass but increases with branching and with the number of hydroxyl (OH) groups.

Procedure: To each of four test tubes, add 10 drops of one of thefollowing alcohols: ethanol, 1-butanol, 2-methyl-2-propanol and ethylene glycol. Add 2 mL of water to each test tube and observe the results.

2. Acid Properties of Alcohols:

Alcohols react with metallic sodium with the evolution of hydrogen. The relative acidities of alcohols and consequently their relative rates of reaction with sodium are in the order: primary > secondary > tertiary.

$$2R-OH + 2Na \longrightarrow 2RO^- Na^+ + H_2$$

Procedure: To each of three test tubes containing a small piece ofsodium, add 2 mL of dry 1-butanol, 2-butanol or 2-methyl-2-propanol. Compare the rates of evolution of hydrogen gas and record your results. After all the sodium has reacted in the test tube containing the 1-butanol, add 3 drops of phenolphthalein indicator solution and observe the colour change.

3. Chromic Acid Oxidation of Alcohols:

Primary and secondary alcohols are oxidized by chromic acid to the corresponding carboxylic acids and ketones respectively. Tertiary alcohols are generally unreactive under similar conditions. When alcohols are oxidized they reduce chromium (VI) to Cr (III) changing the colour of the solution from orange to green. Oxidation therefore offers a method for distinguishing primary and secondary alcoholsfrom tertiary alcohols.

$$K_2Cr_2O_7 + R\ CH_2OH \xrightarrow{H_2SO_4} R-\underset{\underset{O}{\|}}{C}-OH + Cr^{3+}$$

$$K_2Cr_2O_7 + R-\underset{\underset{R}{|}}{\overset{-}{C}}HOH \xrightarrow{H_2SO_4} R-\underset{\underset{O}{\|}}{C}-R + Cr^{3+}$$

(orange)   (green)

Procedure: Place 5 mL of 1% potassium dichromate solution and 2drops of concentrated sulphuric acid in each of three test tubes. Mix thoroughly and add 2 drops of one of the following alcohols: 1-butanol, 2-butanol, or 2-methyl-2-propanol. Shake the tubes and observe any change in colour.

4. *The Lucas Test:*

A solution of zinc chloride in concentrated hydrochloric acid (Lucas reagent) can be used to distinguish between primary, secondary and tertiary alcohols. With this reagent the order of reactivity is typical of compounds reacting by the S_N1 mechanism. The zinc chloride (a Lewis acid) assists in breaking the C-OH bond as illustrated in the equation below:

$$R-OH + ZnCl_2 \rightleftharpoons R-\underset{\underset{H}{|}}{\overset{\delta^+\quad\delta^-}{O}}\cdots ZnCl_2 \xrightarrow{-[HOZnCl_2]^-} R^+ \xrightarrow{+\ Cl^-} R-Cl$$

Alcohols (of no more than six carbons) are soluble in the Lucas reagent while the corresponding alkyl chlorides are not. Tertiary alcohols react rapidly with the reagent forming an insoluble alkyl chloride layer almostimmediately. Secondary alcohols react within 5-10 minutes, while primary alcohols require several hours to react at room temperature(S_{N2} mechanism).

Procedure: Place 2 mL of Lucas' reagent in each of three test tubesand add 6 drops of the alcohol to be tested. Close the tubes with a cork shake and allow them to stand. Record the time required for the tubes to acquire a cloudy appearance. Carry out the test on 1-butanol, 2-butanol, and 2-methyl-2-propanol.

5. *The Iodoform Test:*

This is a test for methyl carbinols having the structure CH_3CHOH-and methyl ketones (CH_3CO-). Methyl carbinols are first oxidized by the reagent to methyl ketones which become iodinated and then cleaved by base to give a bright yellow precipitate of iodoform.

$$R-\underset{\underset{H}{|}}{\overset{\overset{OH}{|}}{C}}-CH_3 \xrightarrow{I_2 / NaOH} R-\overset{\overset{O}{||}}{C}-CI_3 \xrightarrow{OH^-} R-\overset{\overset{O}{||}}{C}-O^- + CHI_3$$

Procedure: To each of three test tubes add 2 mL of 10% NaOHsolution. Add 5 drops of one of the following alcohols: 1-butanol, 2-butanol or 2-methyl-2-propanol. To each test tube add, drop wise with shaking, 5 mL of iodine-KI solution. Observe any changes and record your results.

II. PHENOLS:

The most common reactions of phenols involve breaking the O-H bond and the usual electrophilic aromatic substitution at the aromatic ring. Protonation of the hydroxyl group and loss of a water molecule as in alcohols would give a phenyl cation which is very unstable and difficult to form. Since the aromatic nucleus is electron rich, direct attack by nucleophiles as in S_N^1 or S_N^2 reactions is not possible. Consequently, phenols do not undergo substitution of the hydroxyl group either by the S_N^1 or S_N^2 mechanisms.

The characteristic property that differentiates phenols from alcohols is acidity. Phenols are stronger acids than alcohols and react with sodium hydroxide, whereas alcohols do not. The reason for this difference is that the phenoxide ion is resonance-stabilized whereas the alkoxide ion is not.

EXPERIMENTAL

MATERIALS NEEDED	Glassware: 4 Test tubes.
	Chemicals: cyclohexanol, phenol, *p*-cresol, 4 mL bromine water solution, 0.5 mL ferric chloride solution (1%), 6 mL of 10% NaOH solution .

1. Acidity of Phenols:

Procedure: In each of three test tubes add 0.4 mL or 0.2 g ofcyclohexanol, phenol, or *p*-cresol. Add 1 mL of water to each tube, shake and note whether the compound dissolves. If not add 2 mL of 10% NaOH solution and observe the result.

2. Bromination of Phenols with Bromine Water:

The hydroxyl group strongly activates the aromatic ring towards electrophilic aromatic substitution. Phenol readily forms a tribromo derivative when treated with a solution of bromine-water at room temperature.

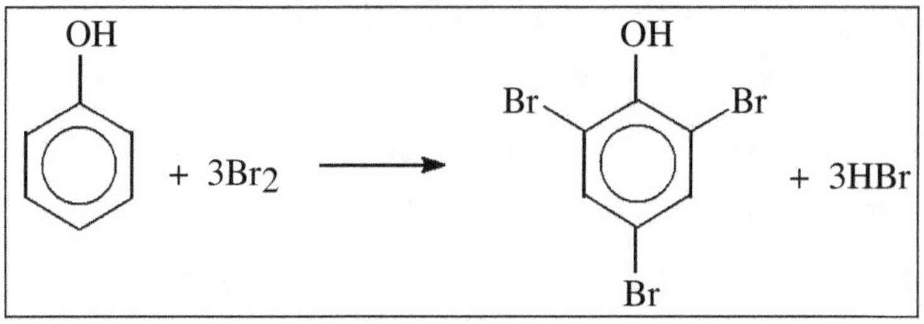

Procedure: In a test tube introduce 1 mL of water and about 0.2 g ofphenol. Add enough bromine-water and shake until the yellow colour persists. Observe the formation of a precipitate.

3. Ferric Chloride Test:

The presence of a phenolic (or enolic group) in a compound is indicated by the formation of a violet (or red) iron complex when treated with a ferric chloride solution

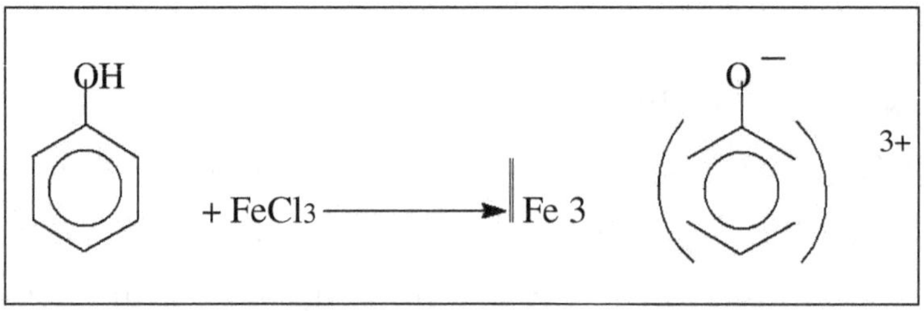

Procedure: In a test tube dissolve a few crystals of phenol in 5 mL ofwater. Add 1-2 drops of 1% ferric chloride solution, shake and observe the results. Repeat with cyclohexanol.

DEHYDRATION OF ALCOHOLS

Preparation of an Alkene and Tests for Unsaturation

INTRODUCTION:

Alcohols having β-hydrogens can be converted into alkenes by dehydration (elimination of a molecule of water). Dehydration of alcohols requires an acid catalyst and heat. Common dehydrating acids such as phosphoric or sulphuric acids are used for this purpose.

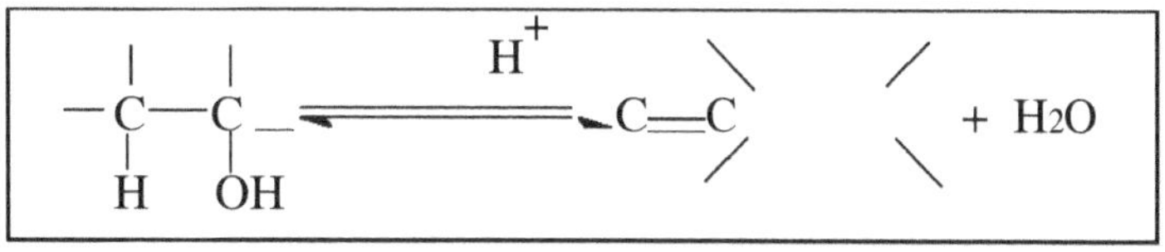

$$-\overset{|}{\underset{H}{C}}-\overset{|}{\underset{OH}{C}}- \xleftarrow{\quad H^+ \quad} \diagdown C = C \diagup + H_2O$$

Since all the steps of the above reaction are reversible, the alkene should be removed from the reaction mixture by distillation as soon as it is formed in order to push the equilibrium forward. Since alkenes are much lower boiling than the parent alcohols, this is experimentally possible.

The various classes of alcohols differ widely in their ease of dehydration: tertiary alcohols dehydrate most easily while primary alcohols are the most difficult to dehydrate. The following examples show how these differences in reactivity affect the experimental conditions of the dehydration, i.e. the temperature and concentration of the acid catalyst employed.

66

In addition to different experimental conditions, the alcohols follow different mechanistic routes. While secondary and tertiary alcohols dehydrate by an E1 mechanism, the dehydration of primary alcohols usually takes place by the E2 mechanism. In either case the first step is protonation of the alcohol. In the E1 mechanism a carbocation intermediate is first formed followed by (in the absence of strong nucleophiles) elimination of an adjacent proton to give the alkene. In the E2 mechanism loss of a molecule of water and elimination of a β-hydrogen occur in one step.

Acid catalysts like H_2SO_4 and H_3PO_4 give, upon dissociation, anions (HSO_4^- and $H_2PO_4^-$) which are weak nucleophiles and therefore minimize substitution products. Acids such as HCl and HBr, on the other hand, are not suitable for dehydration since the nucleophilic Cl^-

or Br⁻ will convert alcohols to alkyl halides instead of to alkenes.

When an alcohol can dehydrate to give more than one alkene, the major product will be the more substituted alkene. i.e. the one with the greater number of alkyl groups attached to the double bond (Saytzeff rule)

$$
\underset{OH}{\overset{CH_3}{CH_3-\underset{|}{\overset{|}{C}}\ CH_2CH_3}} \quad \xrightarrow[100\ ^\circ C]{H_2SO_4} \quad CH_3-\overset{CH_3}{\underset{|}{C}}\ CHCH_3 \ + \ CH_2=\overset{CH_3}{\underset{|}{C}}\ CH_2CH_3
$$

Major   minor

I. PREPARATION OF 2-METHYL-2-BUTENE:

$$
\underset{OH}{\overset{CH_3}{CH_3-\underset{|}{\overset{|}{C}}\ CH_2CH_3}} \quad \xrightarrow[100\ ^\circ C]{H_2SO_4} \quad CH_3-\overset{CH_3}{\underset{}{\overset{|}{C}}}\ CHCH_3
$$

2-methyl-2-butanol   2-methyl-2-butene

(bp 102, den 0.81, mw 88)   (bp 38, den 0.66, mw 70)

EXPERIMENTAL

MATERIALS NEEDED	Glassware: Distillation apparatus with a 150 mL round bottomedflask, separatoryfunnel(50 mL), graduated cylinder (10 mL), test tube, ice bath, water bath. Chemicals: 5 mL Concentratedsulfuric acid, 11 mL 2-methyl-2-butanol, 1 g anhydrous calcium chloride, 10 mL sodium hydroxide (10%).

Prepare a 30% solution of sulphuric acid by adding, in small portions, 5 mL of concentrated sulphuric acid to 10 mL of cold water in a 150 mL round-bottomed flask immersed in a cold-water bath. To this cold solution add in small portions 11 mL (0.10mol) of 2-methyl-2-butanol. Connect a condenser and subject the mixture to distillation using a water bath and an ice-cooled receiver to minimize loss of the highly volatile product through evaporation (Figure 23). Ensure all connections are well sealed to prevent leakage of the flammable alkene. Continue distillation of the crude alkene until the thermometer indicates a rise in temperature above 50°C.

Transfer the cooled product to a small separatory funnel and carefully wash with 10 mL of 10% NaOH solution *(Caution: excess pressuredevelops upon shaking)*. Separate the organic layer and dry it overanhydrous calcium chloride.

Finally distil the dry product over a water bath, collecting the fraction boiling at 35-42°C in a large test tube immersed in

ice. Determine the weight of the product and calculate the yield. Proceed to part III.

II. CYCLOHEXENE FROM CYCLOHEXANOL:

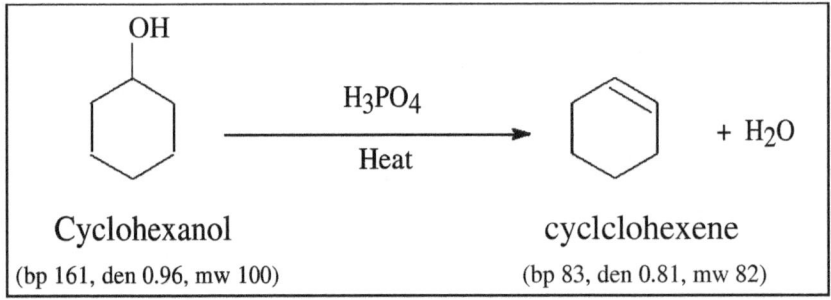

Cyclohexanol   cyclclohexene
(bp 161, den 0.96, mw 100)   (bp 83, den 0.81, mw 82)

In this experiment the secondary alcohol cyclohexanol will be dehydrated to cyclohexene.

69

EXPERIMENTAL

MATERIALS		
	NEEDED	<u>Glassware:</u>Distillation apparatus with a 50 mL round-bottomed flask, separatory funnel(50 mL), 2 test tubes.
		<u>Chemicals:</u>3 mL 85% phosphoric acid, 10 mL cyclohexanol, 1 g anhydrous calcium chloride, 10 mL sodium carbonate(10%).

Place 10 mL (0.1mol) of cyclohexanol in a 50 mL round-bottomed flask. Add 3 mL of 85% phosphoric acid and several boiling chips. Swirl the flask to mix the contents. Arrange for distillation (Figure 23) and heat the mixture gently over a small flame so that cyclohexene distils slowly (along with water) into a test tube immersed in ice and the temperature does not exceed 100. Continue the distillation until about 3 mL of liquid remain in the distillation flask.

Transfer the distillate, which consists of two layers, to a small separatory funnel and extract with 10 mL of 10% sodium carbonate solution. Dry the product over anhydrous calcium chloride. Decant the clear liquid into a 50 mL round-bottomed flask and distil into a dry test tube immersed in ice. Collect the fraction which distils between 79 and 84°C. Cyclohexene is usually obtained as a clear liquid. Determine the weight of your product and calculate its yield. Proceed to part III.

III. TESTS FOR UNSATURATION:
1. Reactivity towards Bromine:
Bromine decolourization is used as a simple test for unsaturation. Bromine is a dark red-brown liquid, while

alkenes and their dibromides are colourless. A dilute solution of bromine in an inert colourless solvent, such as carbon tetrachloride or methylene chloride, is rapidly decolorized when it is added to an alkene. The following addition reaction takes place.

$$-\overset{|}{C}=\overset{|}{C}- \quad + Br_2/CH_2Cl_2 \longrightarrow -\overset{|}{\underset{Br}{C}}-\overset{|}{\underset{Br}{C}}-$$

(colorless)   (red-brown)   (colorless)

2. Reactivity towards Potassium Permanganate:

Another test for unsaturation is the Baeyer test. In this test cold, dilutepermanganate is used to oxidize alkenes to glycol. The reagent is reduced to manganese dioxide which appears as a brown precipitate while the purple colour of potassium permanganate disappears.

$$-\overset{|}{C}=\overset{|}{C}- \quad +2KMnO_4 +4H_2O \longrightarrow -\overset{|}{\underset{OH}{C}}-\overset{|}{\underset{OH}{C}}- \quad +2MnO_2 + 2KOH$$

(purple)   (brown)

EXPERIMENTAL

MATERIALS NEEDED	Glassware: 4 test tubes.
	Chemicals: 2 mL of 1% bromine in methylene chloride, 2 mL methylene chloride, 2 mL potassium permanganate (0.5%). 1mLof alkene.

1. **Bromine Test:** Dissolve 5 drops of the alkene you prepared in 1mL methylene chloride. Add to this solution dropwise1mL of a 1% solution of bromine in

methylene chloride. Observe the result. For comparison repeat the test with an alkane.

Baeyer Test: Add 5 drops of the alkene you prepared to 2 mL ofwater in a test tube. Add 1 mL of 0.5% potassium permanganate solution, shake and note the result. Repeat with an alkane.

QUESTIONS:

1) Dehydration reactions are acid-catalysed. What is the function of the acid in such reactions?
2) Could hydrochloric acid be an acceptable substitute for the acid used in this experiment? Explain.
3) Why was it necessary to wash the crude alkene with a solution of aqueous base?

ALDEHYDES AND KETONES

Classification, Tests and Derivatives

INTRODUCTION:

The carbonyl group is common to both aldehydes and ketones, and as a result, both classes of compounds react similarly with many reagents. 2,4-Dinitrophenylhydrazine is commonly used to test for both types of compounds. However a distinguishing behaviour of aldehydes is their reaction with mild oxidizing agents which oxidize them to carboxylic acids while ketones, which are more difficult to oxidize, remains unchanged.

$$R-\underset{\underset{H}{|}}{\overset{\overset{O}{\|}}{C}} \xrightarrow{\text{oxidizing agent}} R-\underset{}{\overset{\overset{O}{\|}}{C}}-OH$$

$$R-\underset{\underset{R}{|}}{\overset{\overset{O}{\|}}{C}} \xrightarrow{\text{oxidizing agent}} \text{no reaction}$$

Several laboratory tests that distinguish between aldehydes and ketones, therefore, take advantage of this difference in behaviour towards oxidants. One of these is *Tollens'* silver mirror test, in which a silver ammonia complex ion is reduced, by aldehydes, to metallic silver. *Fehling's* and *Benedict's* solutions are also distinguishing reagents where the Cu(II)ion, complexed to tartarate or citrate respectively, is reduced to red cuprous oxide (Cu_2O) by aldehydes but not ketones.

Carbonyl compounds (aldehydes and ketones) are conveniently identified through a number of easily prepared derivatives. These include oximes, phenylhdrazones, 2,4-dinitrophenylhydrazones and semicarbazones. These derivatives are ideal because they are easily purified, crystalline solids with sharp melting points.

The mechanism of formation of these closely related derivatives involves a typical nucleophilic addition at the carbonyl carbon followed by elimination of a water molecule.

or further structural identification of methyl carbonyl compounds, the iodoform reaction, using iodine and aqueous sodium hydroxide is used. Compounds containing the CH_3CO group give a bright yellow precipitate of CHI_3 (*iodoform*).

EXPERIMENTAL

MATERIALS NEEDED	Glassware: 4 test tubes, Erlenmeyer flask (50 mL), ice bath, graduated cylinder (10 mL), Buchner funnel, filter flask, melting point apparatus. Chemicals: 15 mL Tollens' reagent, 0.5 mL each of: formaldehyde, benzaldehyde, acetone, 2-propanol, 2-pentanone, 3-pentanone, 15 mL Fehling's or Bendict's solution, 12 mL sodium hydroxide(5%), 40 mL iodoform reagent, 1.0 g hydroxylamine hydrochloride, 3 g sodium acetate, 2.3 mL cyclohexanone, 30 mL petroleum ether, 5.0 mL phenylhydrazine reagent, 50 mL ethanol, 16 mL 2,4-dinitrophenylhydrazine reagent, 1.0 g semicarbazide hydrochloride and 1.0 g unknown.

I. TESTS AND DERIVATIVES:

1. 2,4-Dinitrophenylhydrazine Test

Procedure: To 2 mL of ethanol in a test tube add 5 drops of acetoneand mix. Add 2 mL of 2 4-dinitrophenylhydrazine reagent shake, and observe the result. Repeat with benzaldehyde and cyclohexanone.

2. Tollens' Silver Mirror Test:

$$R-\overset{\overset{O}{\|}}{C}-H + 2[Ag(NH_3)_2]^+OH^- \underline{\quad\quad} R-\overset{\overset{O}{\|}}{C}-O^-NH_4^+ + 2Ag + 3NH_3 + HO_2$$

Procedure: Place 5 mL of freshly prepared*Tollens'* reagent in each ofthree clean test tubes and add 3-4 drops of formaldehyde benzaldehyde, or acetone solution. Shake the tubes vigorously and allow them to stand for 10 minutes. If no reaction occurs place the tubes in a hot water bath (50^oC) for a few minutes.

3. Fehling's and Benedict's Tests:

$$R-\overset{\overset{O}{\|}}{C}-H + 2[Cu(complex)]^{2+} \longrightarrow R-\overset{\overset{O}{\|}}{C}-O + Cu_2O$$

Procedure: Place 5 mL of either*Benedict's*reagent or freshlyprepared *Fehling's* reagent in a clean test tube and add a few drops of formaldehyde solution. Place the test tube in a beaker of boiling water and observe any colour changes that occur within 15-20 minutes. Record your observations. Repeat with benzaldehyde and acetone.

4. The Iodoform Test:

$$R-\overset{\overset{\displaystyle O}{\|}}{C}-CH_3 + I_2 + OH^- \longrightarrow R-\overset{\overset{\displaystyle O}{\|}}{C}-O^- + CHI_3$$

Procedure: To 10 drops of acetone in 3 mL of 5% sodiumhydroxide add iodine solution and shake until the colour of iodine nearly persists (about 10 mL of iodine solution). At this stage the yellow precipitate of iodoform should have formed. Repeat the test with 2-propanol, 2-pentanone, and 3-pentanone. Record your observations for each compound.

5. CyclohexanoneOxime:

Procedure: Dissolve 1 g of hydroxylamine hydrochloride and 1.5 gof sodium acetate in 4 mL of water in a test tube. Warm the solution to about 40 °C and then add 1 mL of cyclohexanone. Stopper the flask and shake for 1-2 minutes. Cyclohexanoneoxime begins to separate as fine colourless crystals. Cool the tube thoroughly in an ice bath to complete precipitation. Filter the crystals using a small Buchner funnel then wash with a little ice-cold water. Air-dry the crystals and determine their melting point.

Oximes of some carbonyl compounds will crystallize more slowly. They may require longer cooling and scratching the walls of the flask to induce crystallization. Oximes can be recrystallized nicely from petroleum ether (bp 40-50).

6. Cyclohexanone Phenylhydrazone:

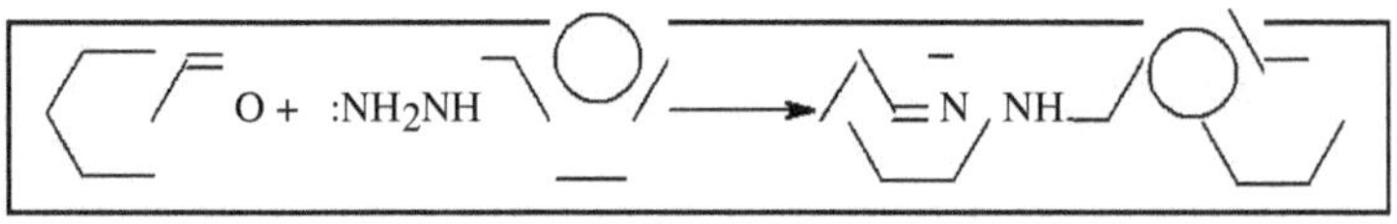

Procedure: Add 10 drops of cyclohexanone to 5 mL of ethanol and 5mL of phenyl hydrazine reagent Shake the mixture until a clear solution is obtained. Heat the mixture on a water bath for 10 minutes then cool in an ice bath and filter the crystals.

The crude phenylhydrazone may be recrystallized by dissolving it in hot ethanol and adding water to the hot solution until a faint turbidity persists. The solution is then cooled and the crystals are collected.

7. Acetone 2,4Dinitrophenylhydrazone

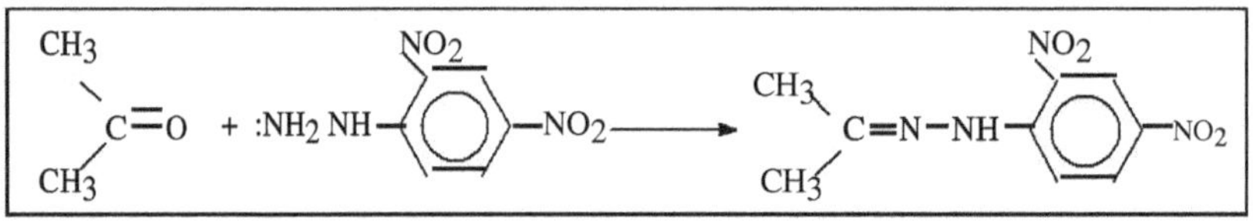

Procedure: Add 10 mL of 2, 4-dinitrophenylhydrazine reagent to 10drops of acetone in 10 mL of ethanol and allow the mixture to stand at room temperature for a few minutes. Precipitation of the acetone 2, 4-dinitrophenylhdrazone usually occurs immediately; other carbonyl compounds may require 5-6 minutes or even several hours.

When crystallization is complete, cool the mixture in an ice bath, and collect the crystals. Recrystallize from ethanol, dry the resulting crystals and determine their melting point.

NOTE: Derivatives of phenylhydrazine are suspected

77

carcinogens.Handle with care and avoid skin contact.

8. CyclohexanoneSemicarbazide:

Procedure: In a test tube dissolve 1g semicarbazide hydrochlorideand 1.5 g sodium acetate in 10 mL of water, add 1 mL of cyclohexanone and shake vigorously. Place the test tube in a beaker of boiling water for 5 minutes then cool to room temperature. Place in an ice bath and scratch the sides of the tube with a glass rod until crystallization is complete. Filter the crystals, wash with a little water, and recrystallize from ethanol. Dry the crystals and determine their melting point.

II. IDENTIFICATION OF AN UNKNOWN CARBONYL COMPOUND

Procedure: While the various derivatives are drying, obtain anunknown from your instructor and proceed to identify it as follows: Use *Tollen's* test to determine whether the compound is an aldehyde or a ketone. If the unknown is a ketone, perform the iodoform test to determine if it is a methyl ketone. Finally prepare a crystalline derivative of the unknown to determine its identity. The unknown is selected from the aldehydes and ketones listed below:

Table 3. Derivatives of some aldehydes and ketones

Compound	Formula	Oxime	Phenyl-hydrzone	2,4-DNP	Semi-carbazone
Ethyl methyl ketone	$CH_3CH_2COCH_3$	oil	oil	116	136
Diethyl ketone	$CH_3CH_2COCH_2CH_3$	69	oil	156	139

Furfural	$C_4H_3O.CHO$	75	79	229;212	202
Crotonaldehyde	$CH_3CH = CHCHO$	119	56	190	199
Benzaldehyde	C_6H_5CHO	35	158	237	224
Cyclohexanone	$C_6H_{10}O$	91	81	162	167
2-Heptanone	$CH_3(CH_2)_4COCH_3$	oil	207	89	127
n-Heptanal	$CH_3(CH_2)_5CHO$	57	oil	108	109
Acetophenone	$C_6H_5COCH_3$	59	105	239	199
2-Octanone	$CH_3(CH_2)_5COCH_3$	oil	oil	58	123
Salicylaldehyde	$C_6H_4(OH)CHO$	63	143	252	231
Cinnamaldehyde	$C_6H_5CH=CHCHO$	64	168	255	215

QUESTIONS:

1) Write a detailed mechanism for the following general reaction for the preparation of derivatives of carbonyl compounds:

$$R_2\text{-}C\text{=}O + NH_2\text{-}G \longrightarrow R_2\text{-}C\text{=}N\text{-}G + H_2O$$

2) Which of the following compounds gives a positive iodoform test?

    1) $CH_3\text{-}CH_2\text{-}CH_2\text{-}CO\text{-}CH_2\text{-}CH_3$

    2) $CH_3\text{-}CH_2\text{-}CH_2\text{-}CH(OH)\text{-}CH_3$

    3) $CH_3\text{-}CH_2\text{-}CH_2\text{-}CH_2\text{-}CHO$

3) With the help of a chemical equation, suggest a role for the acidinthe formation of 2,4-dinitrophenylhydrazones.

THE CHEMISTRY OF AMINES

Classification, Tests and Derivatives

INTRODUCTION;

Amines are basic organic compounds that contain nitrogen and may be considered as derivatives of ammonia. They are classified as primary, secondary, or tertiary, depending on the number of alkyl groups attached to the nitrogen atom.

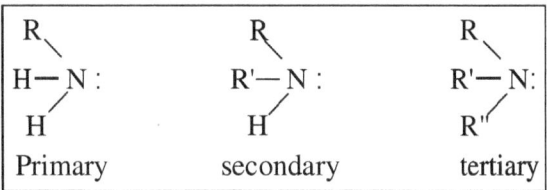

Like ammonia, amines are electron donors and behave both as bases and as nucleophiles. Alkyl amines are considerably stronger bases than aryl amines. The diminished basicity of aryl amines is attributed to the appreciable overlap of the lone pair with the π-electrones of the benzene ring making them less available for donation.

The following tests are intended to illustrate the dual chemical behaviour of amines as both bases and nucleophiles.

EXPERIMENTAL

MATERIALS NEEDED	Glassware: 3Test tubes, Buchner funnel, beaker.
	Chemicals: 0.5 mL n-butylamine, 0.5 mL di-n-butylamine, 0.5 g p-toluidine, 1 mL concentrated hydrochloric acid, 1 mL sodium hydroxide(20%), 2 mL dimethylaniline, 2 mL methyl iodide, 2 mL aniline, 2 mL acetic anhydride 1 mL benzenesulfonyl chloride, 1 mL methylaniline, 4 mL copper sulfate(5%), 15 mL

1. Solubility in Water:

Amines of low molecular weight are water soluble due to their ability to form hydrogen bonds with water. Borderline solubility is reached when the amine has about five carbon atoms.

Procedure: To each of three test tubes add 2 mL of water and 5 dropsof n-butylamine, di-n-butylamine or p-toluidine. Shake the tubes and note the solubility in each case. In the case of soluble amines test the solution with litmus paper.

2. Basicity and Salt Formation:

The basic characteristic of water soluble amines may be shown by the response of their aqueous solutions to litmus paper. Amines which are insoluble in water are readily soluble in aqueous acids forming water-soluble ammonium salts. The salt can be converted back to the amine by making the solution alkaline.

$$RNH_2 \underset{OH^-}{\overset{H_+}{\rightleftharpoons}} RNH_3^+$$

Procedure: To 2 mL of water in a test tube add 0.2 g of p-toluidineand shake, then add a few drops of concentrated HCl and note the result. Add enough 20% NaOH until the solution becomes alkaline to litmus and observe the result.

3. Quaternary Ammonium Salts:

Tertiary amines react with alkyl halides to form quaternary ammonium salts.

$$\overset{\backslash}{\underset{/}{\underline{}}} N:+ \ R \ -X \ \longrightarrow \ \overset{\backslash}{\underset{/}{\underline{}}} \overset{+}{N}-R \ + \ X^{-}$$

quaternary ammonium salt

Procedure: Place 1 mL of dimethylaniline in a dry test tube and addabout 2 mL of methyl iodide. Stopper the tube and allow to stand with occasional shaking for 15-20 minutes. Note the formation of a crystalline product. Remove some of the crystals with a spatula and test their solubility in water.

4. *Amides*:

Acetyl derivatives of primary and secondary amines (*acetamides*) are readily prepared by reaction of the amine with acetyl chloride or acetic anhydride.

$$R-NH_2 + CH_3-\overset{O}{\overset{\|}{C}}-X \longrightarrow R-\overset{H}{N}-\overset{O}{\overset{\|}{C}}-CH_3 + HX$$
$$(X = Cl, CH_3COO)$$

Procedure: To 1 mL of aniline in a dry test tube cautiously add 2 mLacetic anhydride. Allow 2-3 minutes for completion of the reaction then cool and collect the solid product by filtration. Test its solubility in dilute hydrochloric acid and compare with that of aniline.

4. *Sulphonamides (The Hinsberg Test):*

Amines react with benzenesulfonyl chloride to give sulphonamides. Sulphonamides derived from primary amines are acidic and dissolve in alkali, while those derived from secondary amines are insoluble in alkali.

Tertiary amines do not react with the reagent. This reaction is used in the *Hinsberg Test* to distinguish among primary, secondary and tertiary amines.

Procedure: To 6 drops of an amine in a test tube add 5 mL of 10%NaOH solution and 8 drops of benzenesulfonyl chloride. Stopper the tube and shake the mixture vigorously until all the benzenesulfonyl chloride has reacted. Cool the solution and decant from any residue. Note whether the residue is solid or liquid and test its solubility indilute HCl. Perform the test on each of the following amines: aniline, methyl aniline and dimethyl aniline.

6. Copper Sulphate Test:

Amines give characteristic, coloured complexes, with copper sulphate.

This simple test is used for the characterization of amines.

Procedure: To 2 mL of 5% copper sulphate solution in a test tube add afew drops of an amine, shake and observe the colour change. Perform the test on aniline and n-butyl amine.

ESTERIFICATION OF ALCOHOLS AND PHENOLS

I. *PREPARATION OF ASPIRIN FROM SALICYLIC ACID:*

Aspirin (*acetylsalicylic acid*) is a widely used drug in modern society. It is an analgesic (pain killer), a powerful antipyretic (fever-reducing) and an anti-inflammatory (swelling-reducing) substance.

Salicylic acid (which is a constituent of certain plants) is itself an analgesic and was originally administered as sodium salicylate. Since salicylic acid has an irritating effect on the stomach lining, chemists sought a modification which would retain its properties while decreasing the adverse side effects. Conversion to the ester satisfied this requirement and acetylsalicylic acid (*aspirin*) proved to be as effective as sodium salicylate without the irritation typical of phenolic compounds. Aspirin gets, however, hydrolysed to salicylic acid in the alkaline medium of the intestines.

Aspirin is obtained from salicylic acid by acetylation with acetic anhydride in the presence of sulfuric acid as catalyst. Because of its low solubility in water (0.25 g/ 100 mL) it is isolated from the reaction mixture by precipitation with water.

EXPERIMENTAL

MATERIALS NEEDED	Glassware: 100 mL Erlenmeyer flask, Buchner funnel, filter flask, graduated cylinder (10 mL), 2 test tubes.
	Chemicals: 3.0 g Salicylic acid, 6 mL acetic anhydride, 1 mL concentrated sulfuric acid, ice, 0.5 mL ferric chloride solution(1%).

salicylic acid   acetic anhydride   acetylsalicylic acid
(mp 159, mw 138)   (bp 137, mw 102)   (mp 128, mw 180)

Place 3.0 g (0.02 mol) of salicylic acid in a 100 mL Erlenmeyer flask and add, with constant stirring, 6 mL of acetic anhydride (fume hood) followed by 1 mL of concentrated sulphuric acid. Stir the mixture gently observing the rise in temperature to 70-80°C while the salicylic acid dissolves. After 15 minutes the solution cools by itself to 35-45°C and a solid mass of aspirin forms. Pour 35 mL of ice-cold water over the contents of the flask to hydrolyse excess acetic anhydride and to complete the precipitation of aspirin. Collect the crude aspirin using a Buchner funnel and wash with ice-cold water. Air-dry the product weight, and calculate the yield.

Ferric Chloride Test: The presence of a phenolic (or enolic) groupin a compound is indicated by the formation of a violet (or red) iron complex when treated with ferric chloride solution.

Procedure: To each of two test tubes containing 1 mL of water, add afew crystals of salicylic acid or aspirin. Now add 1-2 drops of 1% ferric chloride solution, shake and observe the results.

II. PREPARATION OF METHYL BENZOATE:

Carboxylic acids react with alcohols to produce esters in a reaction known as the Fischer esterification. Sulphuric acid is used as a catalyst. Since the reaction is reversible the law of mass action applies.

$$RCOOH + R'OH \rightleftharpoons RCOOR' + H_2O$$

$$K = \frac{[\text{ester}][\text{water}]}{[\text{acid}][\text{alcohol}]} = 4$$

When equal concentrations of alcohol and carboxylic acid are used, the maximum yield of the ester, under equilibrium conditions, is about 67 per cent. It may be increased to nearly 100% either by use of a large excess of one of the reactants (usually the alcohol) or by removal of the water as fast as it is formed.

Mineral acids catalyse the reaction by protonating the carbonyl oxygen of the organic acid (first step) thus making it more susceptible to nucleophilic attack by the alcohol. Furthermore, the acid catalyses the removal of a water molecule (last step).

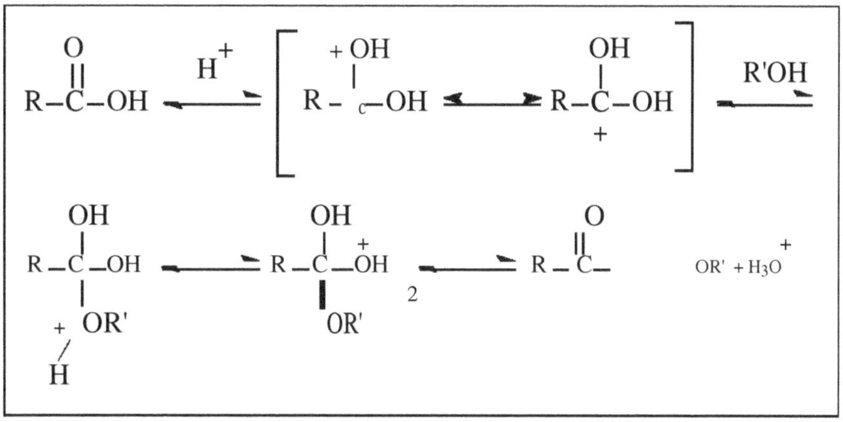

In this experiment, methyl benzoate will be prepared by the sulphuric acid-catalysed esterification of benzoic acid with methanol.

EXPERIMENTAL

MATERIALS NEEDED	Glassware: round-bottomed flask (100, 150 mL), separatory funnel, distillation appartus, 2 Erlenmeyer flasks (50 mL), 2 stands, wire gauze, 2 clamps, 2 clamp holders, graduated cylinder (10 mL).
	Chemicals: 6.1 g Benzoic acid, 20 mL methanol, 1.5 mL concentrated sulfuric acid, 35 mL dichloromethane, 30 mL sodium carbonate (5%), 1 g anhydrous magnesium sulphate, ice..

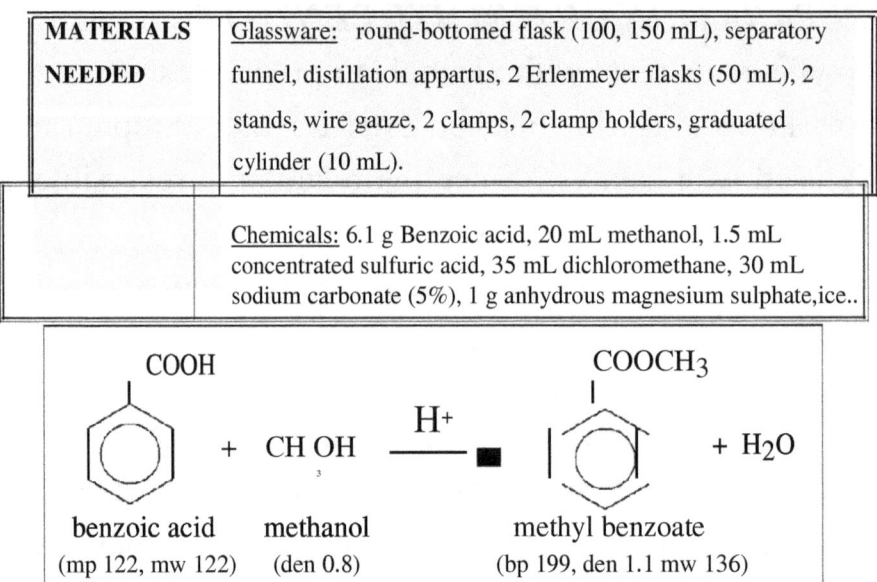

In a 150 mL round bottomed flask place 6.1g of benzoic acid and 20 mL of methanol. Carefully pour 1.5 mL of concentrated sulphuric acid down the walls of the flask and swirl to obtain good mixing. Add two small boiling stones, attach a condenser, and reflux for 1 hr.

Cool the solution to room temperature, then pour into a separatory funnel containing about 25 mL of water and 25 mL of dichloromethane. Rinse the reaction flask with 10 mL of dichloromethane and pour this into the separatory funnel. Shake the mixture vigorously and separate the organic layer from the aqueous layer which contains the excess sulphuric acid and methanol. Wash the organic layer first with 15 mL of water, then twice with 15 mL of 5% aqueous sodium carbonate (*caution: foaming!, shake gently*) to remove unchanged benzoic acid. Carefully separate the organic layer from the aqueous layer and finally wash it with 10 mL of water.

Place the organic layer in a dry flask, add anhydrous magnesium sulphate and cork the flask firmly. Shake thoroughly, and allow the mixture to stand with occasional shaking until it becomes clear. Decant the liquid into a dry round-bottomed flask and distil off the dichloromethane using a water bath and ice-cooled receiver. When no more solvent distils over, continue the distillation using an air-cooled condenser and a direct flame. Collect the fraction boiling at 190-196°C in a dry flask. Determine the weight and calculate the yield.

QUESTIONS:

1) What is the function of the acid in the preparation of methyl benzoate? (give equations).
2) What experimental means may be used to drive an esterification towards completion?
3) Tell how a mixture of benzoic acid and methyl benzoate may be separated?
4) What impurities are most likely to be present in the aspirin you prepared?
   Show (by equations) how they are formed.

p-BROMOANILINE FROM ANILINE

Acetylation, Bromination and Hydrolysis Reactions:

INTRODUCTION:

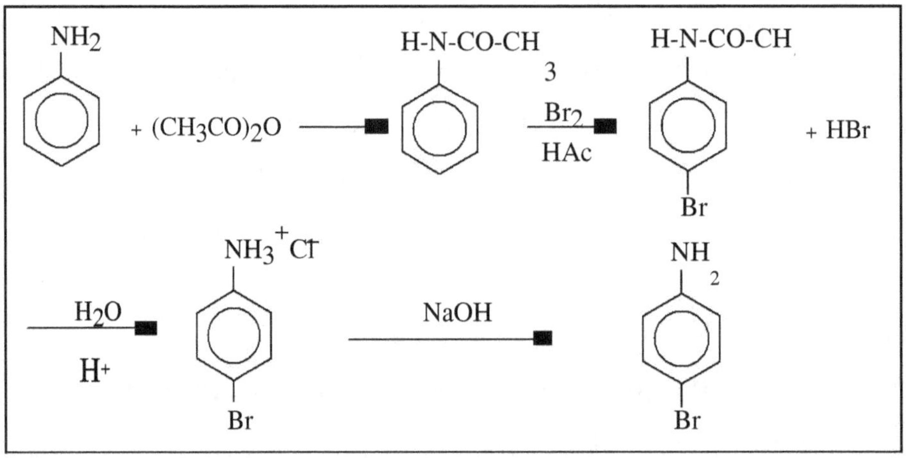

This experiment illustrates the common practice of protecting an active functional group, carrying out the desired reaction (bromination in this case) then removing the protecting group (deprotection).

The amino group is a strongly activating ortho, para-directing group which upon bromination of aniline would give the 2,4,6-tribromo derivative. When monobromination is desired the amino group is converted to the less activating acetamido ($-NHCOCH_3$) group. This bulky group further serves to inhibit formation of the ortho-isomer, which is therefore only a minor by-product in the bromination of acetanilide. Once bromination is carried out, deacetylation of p-bromoacetanilide is achieved by mild hydrolysis.

	EXPERIMENTAL
MATERIALS NEEDED	**Glassware:** beakers (150 and 250 mL), Erlenmeyer flask (100 mL), Buchner funnel, thermometer, graduated cylinder (10 mL), distillation apparatus with 100 and 250 mL round-bottomed flasks, filter paper, litmus paper. **Chemicals:** 4.7 mL aniline, 43 mL glacial acetic acid, 8 mL
	acetic anhydride, 2.8 mL bromine, about 1 g sodium sulfite, 25 mL concentrated hydrochloric acid, sodium hydroxide solution (20%), alcohol for recrystallization, ice.

PREPARATION OF ACETANILIDE:

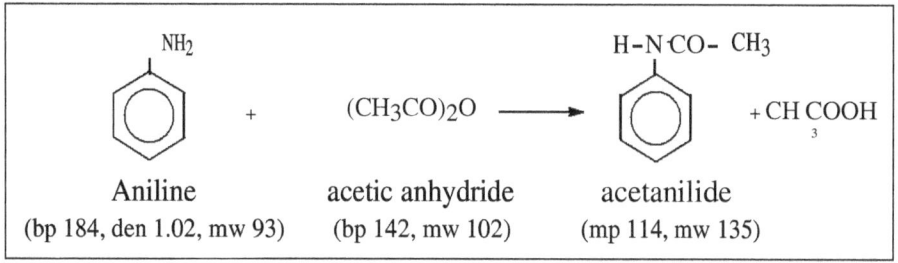

Aniline
(bp 184, den 1.02, mw 93)

acetic anhydride
(bp 142, mw 102)

acetanilide
(mp 114, mw 135)

Place 4.7 mL (0.05mol) of aniline, 8 mL of glacial acetic acid, and 8 mL of acetic anhydride into a 100 mL round-bottomed flask. Attach a reflux condenser and heat the solution to boiling for 10 minutes using a direct flame. Cool the flask under the water tap and pour its contents into a beaker containing 30 mL of water and 30 gof ice and mix thoroughly to enhance precipitation. Collect the crystals of acetanilide on a Buchner funnel, rinse with a little iced water, dry the crystals and use in the next step.

BROMINATION OF ACETANILIDE:

acetanilide (mp 114, mw 135) + Bromine (den 3.1, mw 160) $\xrightarrow{CH_3COOH}$ p-bromoacetanilide (mp 164, mw 214) + HBr

Dissolve 6.75 g (0.05 mol) of acetanilide in 25 mL of glacial acetic acid in a 100 mL Erlenmeyer flask. Warm the flask gently over aflame to assist dissolution. Ensure the temperature of the flask is below 40°C before adding, in small portions and with constant swirling, a solution of 2.8 mL (0.05 mol) of bromine dissolved in 10 mL of glacial acetic acid (fume hood). Within 5-10 minutes bromination is complete and the red colour of the bromine disappears.

The reaction mixture is poured into a beaker containing about 150 mL of cold water and a little ice. If the solution is yellow or red due to excess bromine, add with stirring enough sodium sulphite to discharge the colour. Collect the white p-bromoacetanilide on a Buchner funnel, wash with cold water and, if time permits, recrystallize a small portion from ethanol/water. Any ortho-isomer produced is more soluble and remains in the mother liquor.

HYDROLYSIS OF p-BROMOACETANILIDE:

p-bromoacetanilide (mp 164, mw 214) + H_2O $\xrightarrow{HCl}$ $NH_3^+Cl^-$ $\xrightarrow{NaOH}$ p-bromoaniline (mp 64, mw 172) + HBr

Transfer the crude p-bromoacetanilide into a 250 mL round- bottomed flask and add 25 mL of water and 25 mL of concentrated hydrochloric acid. Heat the solution under reflux for 15 minutes. Cool the flask and add sufficient 20% sodium hydroxide solution to make the solution basic to litmus and to precipitate the p-bromoaniline. Attach the flask to a distillation set-up and steam distil the p-bromoaniline from the aqueous reaction mixture. If the p-bromoaniline solidifies in the condenser, temporarily stop the water flow until the solid melts. Cool the distillate and collect the solidified p-bromoaniline on a Buchnerfunnel and dry between sheets of filter paper. Recrystallization may be performed using ethanol. Weigh the product, determine its melting point, and calculate the percentage yield.

PREPARATION OF CARBOXYLIC ACIDS

Preparation of Benzoic Acid

INTRODUCTION:

Aliphatic and aromatic carboxylic acids can be prepared by a number of methods as shown below. Some methods maintain the same number of carbon atoms in the product, others extend the chain by one carbon atom, while others still, degrade the carbon chain.

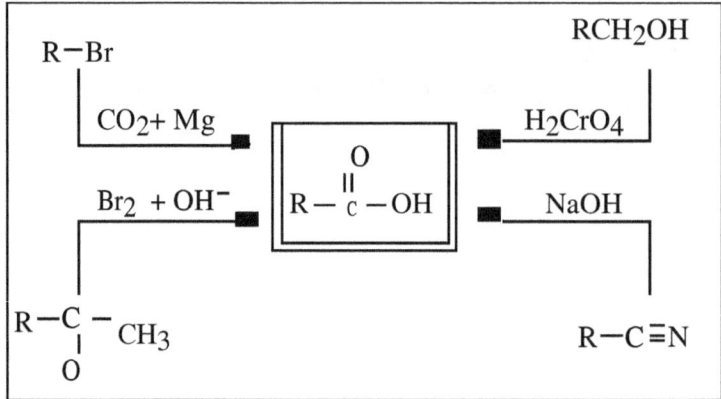

Aromatic acids are also accessible by vigorous oxidation of aromatic side chains.

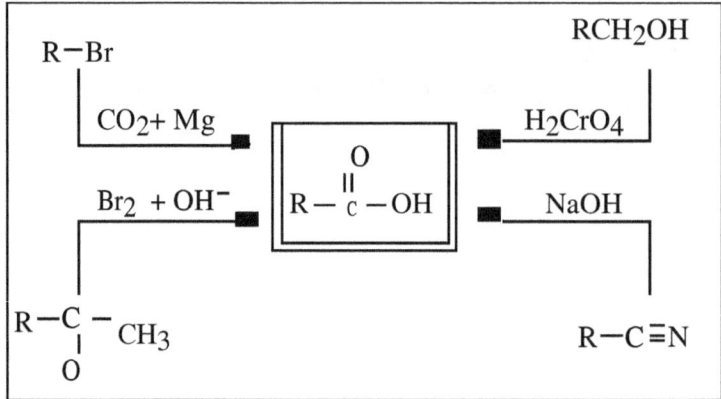

I. BENZOIC ACID BY HYDROLYSIS OF BENZONITRILE:

Acid or base-catalysed hydrolysis of nitriles proceeds in two stages: the amide is formed first and further hydrolysis converts it to the carboxylic acid.

$$R—C \equiv N \xrightarrow[\text{H}^+ \text{ or OH}^-]{\text{H}_2\text{O}} R—\overset{\overset{\text{O}}{\|}}{C}—NH_2 \xrightarrow[\text{H}^+ \text{ or OH}^-]{\text{H}_2\text{O}} R—\overset{\overset{\text{O}}{\|}}{C}—OH + NH_3$$

Under basic conditions the acid gets converted to the water-soluble salt from which the free acid may be regenerated by acidification with a mineral acid.

$$R—C \equiv N \xrightarrow{\text{H}_2\text{O}/\text{OH}^-} R—\overset{\overset{\text{O}}{\|}}{C}—O^- \text{Na}^+ + NH_3 \xrightarrow{\text{HCl}} R—\overset{\overset{\text{O}}{\|}}{C}—OH + NaCl$$

EXPERIMENTAL

MATERIALS NEEDED	Glassware: Round-bottomed flask (100 mL), condenser, Buchner funnel, filter flask, melting point apparatus, ice bath. Chemicals: 2mL benzonitrile, 25 mL sodium hydroxide (10%), 5 mL concentrated hydrochloric acid.

Benzonitrile (den 1.0, mw 103) — CN $\xrightarrow{\text{NaOH}}$ — COO⁻Na⁺ $\xrightarrow{\text{HCl}}$ — COOH benzoic acid (mp 120, mw 122)

NOTE: Benzonitrile is toxic, avoid skin contact.

Assemble a reflux apparatus using a 100 mL round-bottomed flask. Place 2 mL of benzonitrile, 25 mL of 10% sodium hydroxide solution, and two boiling stones in the flask and reflux the mixture gently until the oily layer of benzonitrile disappears (40-50 minutes). Ammonia may be detected during reflux by holding a wetted piece of red litmus paper to the top of the condenser. Cool the mixture tohydrochloric acid with constant stirring; until the contents are acidic to litmus and precipitation of benzoic acid is complete.Slowly

94

add concentratedroomtemperatureandtheninanicebath.

Collect the benzoic acid on a Buchner funnel and wash the precipitate with 5 mL of ice-cold water. Recrystallize the crude product from boiling water. Air-dry the pure benzoic acid then weigh and determine the melting point.

II. BENZOIC ACID BY THE HALOFORM REACTION:

A characteristic reaction of methyl ketones is their conversion to carboxylic acids and halo form upon treatment with halogen and alkali *(Haloformreaction)*. In this experiment benzoic acid is prepared from acetophenoneby a degradation reaction using bromine and sodium hydroxide. The haloform reaction is often used as a test for the presence of $CH_3CH(OH)$ and CH_3CO groups.

EXPERIMENTAL

MATERIALS NEEDED	Glassware:2 Erlenmeyer flasks (250 mL), separatory funnel, Buchner funnel, filter funnel, filter flask. Chemicals: 2 mL Acetophenone, 2mL bromine, 4 g sodium hydroxide, 1 g charcoal, 5 mL concentrated sulfuric acid, 1 g sodium metabisulfide, litmus paper.

Acetophenone
(bp 202, den 1.0, mw 120)

benzoic acid
(mp 122, mw 122)

In a 250 mL Erlenmeyer flask, dissolve 4.0g (0.1 mol) sodium hydroxide in 50 mL of water then add about 50 g of crushed ice. In the fume hood, introduce 2 mL of bromine into the flask and swirl the contents until the bromine dissolves. Add 2 mL of acetophenone and shake well for 10 minutes upon which bromoform forms and settles to the bottom of the flask as a dense liquid. Separate the bromoform in a separatory funnel and pour it into the residue bottle.

The aqueous layer which may be turbid due to residual bromoform is transferred into a 250 mL flask and stirred with 1g of charcoal to remove suspended bromoform. Filter the mixture and cool the filtrate by the addition of ice. Acidify the ice-cold solution with concentrated sulphuric acid until acidic to litmus. Add a little sodium metabisulphite to discharge any orange colour that may develop due to excess bromine. Collect the benzoic acid by suction filtration and recrystallize from the minimum volume of boiling water. Determine the melting point and the percentage yield

QUESTIONS:

1) Explain with the help of equations the function of the base in the hydrolysis of benzonitrile.
2) What impurity might be found in the benzoic acid prepared from benzonitrile? How does this impurity arise?
3) Would it be possible to use the nitrile method to achieve the following conversions? Illustrate with equations.
   a) Ph-Br $\longrightarrow$ Ph-COOH
   b) Ph-CH$_2$Br $\longrightarrow$ Ph-COOH
   c) t-Bu-Br $\longrightarrow$ t-Bu-COOH

II. Benzoic Acid by the Haloform Reaction:

Write an equation for the preparation of benzoic acid from acetophenone:

Theoretical yield of benzoic acid:

Actual yield:

Percentage yield:

Melting point of benzoic acid:

QUESTIONS:

1) Which of the following compounds will give the haloformreaction?

    a)CH_3COOH

    b) $C_6H_5CH_2-CO-CH_3$

    c)$C_6H_5CH(OH) CH_3$

    d) $C_6H_5CH_2-CO-OCH_3$

    e)$C_6H_5-CO-CBr_3$

    f) $C_6H_5CH(OH)CH_2CH_3$

2. Write an equation for the reaction of bromine water with aniline

3. Write equations to show how p-bromoaniline may be prepared starting with bromobenzene.

4. Write the formula and give the name of the compound that may be hydrolysed to formic acid (1 mole) and aniline (1 mole).

5. Explain why the mixture was made alkaline before carrying out steam distillation.